PROFESSIONAL

I0071574

CITIZEN MANUAL 1
INDIVIDUAL TACTICAL SKILLS
FOR THE
PROFESSIONAL CITIZEN

Developed by
The Professional Citizen Project

www.TPCproject.com

For Virginia

Citizen Manual 1 (CM-1)
Individual Tactical Skills

Copyright © 2023 by The Professional Citizen Project

ISBN: 979-8-218-95637-0

Printed in the USA

www.TPCproject.com

CITIZEN MANUAL 1
INDIVIDUAL TACTICAL SKILLS

Contents   Page

Contents

INTRODUCTION

Why the Professional Citizen Series of References?

Sorting through the massive amount of tactics and preparedness information is overwhelming. There has been a trend for years on social media and video hosting services to provide an indiscernible mix of relevant and irrelevant information to Citizens. This content had components of absurd and unrealistic levels of expectation right alongside valid, usable info. It has been an embarrassment of riches; a mountain of information that could not be distilled down to something usable. People seeking echo-chamber validation for their purchases and other folks with a legitimate desire to learn are both trying to drink at the same watering hole. Content watchers (and even students at some classes) were fed an unending stream of shooters banging away at 7 yards with magazine fed carbines and iterations of team CQB with flashbangs. But the normal US dad, mom, wife, husband, brother, son etc were (mostly) treated as an afterthought, left with no real way to sort through the mess and find a starting point. Available print references were in a similar state, many focused on executing tasks with a government level equipped force...or digging a bunker out in the back 40. There were alternatives, however none were squarely focused on translating military tactics and techniques directly to applicable Citizen requirements.

This series of manuals was developed from the need provides you with a Citizen centric solution; a comprehensive series of references that is built specifically for your use. The Professional Citizen series of references is a focused set of references for you to build and grow your tactical skills.

What is the Professional Citizen Reference Series?

The Professional Citizen series of references is a clearly written, easy to use set of references that address baseline individual tasks through complex small unit tactics - all done from the Citizen perspective. This set of references is not an Army or USMC manual; these publications are written to be Citizen specific. We have adapted relevant tactics, doctrine, and best practices for the Citizen and packaged them into a practical series of references. The Professional Citizen series references are written by subject matter experts from a wide variety of backgrounds including military, outdoorsmen, and industry professionals with decades of combined experience.

The Professional Citizen must train and execute in a resource constrained environment. Our community uses US military tactics as the starting point because they are proven and are the most available resource for us to reference. The challenge with using these military sources for our community is the system is designed for a non-resource constrained organization (the military). The methods and the resources far outstretch what we as Pro Citizens can muster. This is the challenge with current materials - finding relevant tactical and doctrinal content designed *specifically for the "average" Citizen* was nearly impossible. The Professional Citizen series of manuals does not include how to conduct air assaults, load helicopters, or emplace cratering charges. The CM manuals provide things that are relevant to you; referencing only those tasks directly related to anticipated Citizen missions.

This series of manuals is grounded in proven Tactics, Techniques, and Procedures (TTPs) and doctrine. This series of references is not a renamed Army manual with a new cover or a repackaged compilation of existing manuals. The references provide you, the Pro Citizen, with **applicable** knowledge and skill development. Yes, some content is a direct lift from current doctrine because a

2

particular subject or reference does not require adjustment to meet your requirements. These references are designed to provide the technical and tactical data in a manner that enables you to select the material that best suits your requirements. Everything from basic individual tasks through planning and leading organizations during a crisis is contained in the series.

Crawl, Walk, Run

The *"crawl, walk, run"* methodology for developing a fighting man (and leaders) is a simple but proven method. The crawl phase can apply to all members of the community no matter how experienced.

We are all new at everything at one point in our journey. Don't allow yourself to be gatekept or tacti-shamed into not improving...or even starting.

Building initial skills and proficiency is critical before moving on to more complex tasks. Learning a new foundational task or a set of tasks sets the bedrock to build on. You will start to add more complex tasks that these basic tasks support (referred to as nesting). As these come together you will build proficiency and credibility as a team member. Be realistic, be deliberate, and be patient with yourself as you build expertise. These references are organized so you can choose your entry point and select the skills that *you* require - from basic to complex. Skill levels vary among us, we all enter the process at different points on the spectrum. Building skills through this deliberate approach will ensure you have not missed a foundational / fundamental element of your training along the way.

This Manual, CM-1, Individual Tactical Skills

The audience for this series of references is the Pro Citizen with a commitment to excellence who needs a framework to help prepare and train. CM-1 is the starting point (the crawl phase) for most as it contains overarching concepts and basic tactical and technical skills that are the bedrock to build on. Mastering the tasks in this reference builds a foundation for you to continue to develop and train more complex skillsets or can assist in sharpening your approach as a seasoned tactician or prepper. Understand that this manual is just a starting point that provides some of the necessary foundational competences to build upon with the skills and tasks contained in the other Professional Citizen series manuals and guides.

THE PROFESSIONAL CITIZEN PROJECT

If not you, who will?

CHAPTER 1
The Professional Citizen

What is a Professional Citizen?

The Professional Citizen is an individual of *any* age, religion, race, either gender, experience, or socio-economic status who is determined to be an active participant in solving future problems for their family and their community. The Pro Citizen is able and willing to assist law enforcement, military, local government and their community during times of crisis. Professional Citizens are average everyday people in the community; they may even be current military, LE, or first responders. The attribute that binds us is we have a mindset of preparedness and service to not only our families but to a cause greater than ourselves. He or she is fully vested in supporting the Natural Rights of Man outlined in the Constitution of the United States or the rights protection documents of their home country. The Pro Citizen has the drive to acquire the skills to Shoot, Move, Communicate, and Survive in any scenario. They take this requirement seriously; being a Professional Citizen is not a hobby. It is a mindset, a culture, and a commitment.

Attributes

The Professional Citizen (Pro Citizen) values preparedness, physical fitness, training, and prides themselves on being proficient with all the relevant tools at their disposal. He has developed and practiced critical thinking skills to

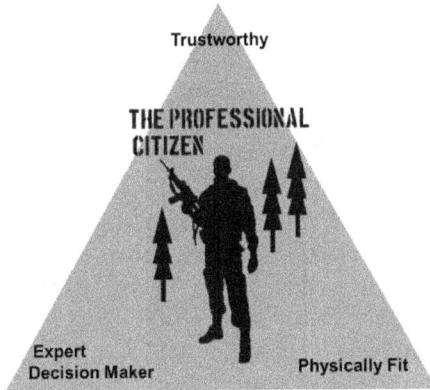

Trustworthy

THE PROFESSIONAL CITIZEN

Expert Decision Maker

Physically Fit

make decisions under extreme pressure with minimal information and time. He has requisite skills that will be called upon in a crisis and has placed value on imparting that knowledge and mindset to his family, group, and community. He takes this responsibility seriously and builds his life around it; it is a way of life. This is a lifestyle that does not have to consume all your resources, the principles can be followed and implemented while maintaining a very "normal" financial life. No need to dig the bunker, max out the credit card on number 10 cans of freeze dried food, or go in debt to buy pano night vision. The Pro Citizen is self-aware and knows his blind spots; he seeks out training and self-improvement to fill in those knowledge gaps. He is a member of a community and is part of a team; either as a member, organizer, or as a leader. He trains individually and as part of that team. He does not have a lone wolf mentality, nor does he see his role as only active if there is a widespread kinetic event. Power outages, adverse weather events, any disruptions of normality that impact his community members...all of these are events when he will step up to help his neighbors out.

Trust. Trust is the foundation for any great organization. We can train members past almost any deficiency if they are grounded in traditional values and morals, honorable, have integrity, and can be counted on when things are hard. Groups can train you to be a good decision maker and can even assist with overcoming physical fitness shortfalls, but integrity and morals must be ingrained before the process. When selecting a team member this must be the absolute starting point. Sometimes trust issues may not become apparent until inside a training cycle so look for indicators early and often. Immediately remove a member that shows a lack of integrity. Being honest and owning mistakes during training, following through on commitments, and treating other members with respect (the Golden Rule) are all indicators of trust. You must be the person that can be counted on to drive on and perform when they are cold, wet, and tired. Your team and family need to know that you will be unwaveringly reliable.

Expert Decision Maker. Decision making is the cornerstone of all tactical operations. From micro decisions that must be made constantly to complex mission planning and execution. The ability to make sound decisions in a timely manner can be trained and developed, but there is a "common sense" starting point that must be present in a team member. From the point man who must make split second decisions to the patrol leader who must plan and adjust a patrol base location, all members must be capable of making good decisions. Not everyone is cut out to be a leader, but everyone must be able to make decisions and act at some level. Professional Citizens have to make decisions that will affect not only them but the entire team. Being a person who can do so is an incredible asset for the group. The follow on to this concept is leaders for the organization will come from within; selecting a new team member is a vote for future consideration in a leadership position. Leader development is critical for the organization, identifying and developing those with leadership potential will be a constant effort.

Physical Fitness. Why would a Tactical Skills manual address physical fitness? Look around any public place and count the obese. As a nation we are extremely fat and unfit for combat. The US rapidly became a fast food nation of overweight people due to several factors, the greatest among them being the typical poor quality diet of an American. Out of shape people are combat ineffective before the fight ever starts, we must start changing that time now.

When it comes to the Citizens that will be called upon to go out and recon, ambush or raid, these combatants must be true fighting men and women. They do not have to be young or 8 percent bodyfat fitness gurus...but they must be physically able to shoot, move, communicate, and survive for days or weeks at a time. They are the main effort and are always the priority for community resources; if they fail the entire community is done for. There is a considerable physical fitness obligation that comes with this role and responsibility, one that cannot be trained or added after X Hour. Being as physically fit as possible while filling any function in the community is critical. Fit people stay healthier, can stay awake longer, work harder, and will perform better under stress. If you get anything out of this manual this is it. Put a personal nutrition and fitness plan in place and get started. *Consult with your physician before starting any fitness regimen to ensure your baseline health poses no risk to doing so.* Starting can be as simple as cutting out fast food and sodas and walking 30 minutes a day. If you are just starting your journey, begin there and build upon it through a comprehensive fitness program.

According to the CDC:

"in the early 1960s, fewer than 14 percent of the individuals possessed a body mass index (BMI) of over 30. Today, the figure collected by the CDCs National Health and Nutrition Examination Survey (NHNES) is 43 percent."

We are combat ineffective as a nation

The flip side of this same coin is a contributing member does not necessarily need to be a lean, mean, fightin' machine or be a direct combatant to have a vital role in the effort. A contributing member of the preparedness community needs only to have the mindset, the "heart" if you will, to be an asset for the group. Finding and emphasizing strengths in the community is where it's at. The 80 year old grandfather who was a career electrical engineer and ham radio enthusiast, the obese dude that used to be a Navy cook and now owns a food truck, and the mailman who has driven and knows every family in the community are all assets. There is talent in almost (*almost*) every person; identify, recognize, and leverage it. Everyone must still have a baseline of skills that can be called upon even if not physically capable of rucking up and defending against threats. Grandma must still know how to safely load, clear, and reduce a malfunction on a striker fired pistol, an M4 or an AK. But the Professional Citizen, the one who is called upon when a threat is upon us - they must be fit to fight.

Road to War (The Process)

Decision. Some in the community have been all about it for years, some are just starting out, and some have not yet seen the light. For one reason or another you have made the decision to pick this manual up and start getting after it or you may have selected it to enhance your ongoing training and preparation efforts. Every single day counts. Our team is not comprised of alarmists or doomers, we are simply realistic optimists with a sense of urgency. The one thing we cannot choose is when and where an event may occur, so the best way to ensure positive outcomes in our future is to be ready *before* those bad things happen.

Professional Citizen Road to War

- Events may occur in parallel
- Training Individual Tasks as a group is optimal
- "War" can mean anything from localized short-duration emergencies to global thermonuclear war

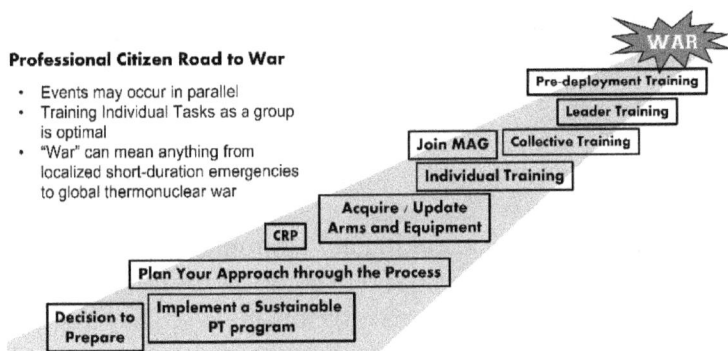

WAR

Pre-deployment Training

Leader Training

Join MAG | Collective Training

Individual Training

Acquire / Update Arms and Equipment

CRP

Plan Your Approach through the Process

Decision to Prepare | Implement a Sustainable PT program

The Road To War

Plan Your Approach Through the Process. The process depicted in the figure above is simply a framework to follow. It is a roadmap of the big pieces to be planned for and accomplished as soon as possible. The term "Road to War" is simply context to model a deliberate approach to preparing. "War" may simply be a local weather event or civil unrest that disrupts you and your family. It does not necessarily mean global conflict or a foreign invasion; it is just a focus point to get us moving in the right direction for training and material preparations. H Hour is a military term for when an operation kicks off, for our purposes we will refer to the crisis event start as X Hour in this manual. This is when "it" starts and there will be no further opportunity to prepare.

Allocate some quiet time to lay out your way ahead. Capturing events or block times on calendars is the best way to visualize the process and it becomes a promise to yourself. Some blocks are sequential, some can be done in parallel to save time. It is a significant undertaking, attacking things a small chunk at a time can make this manageable.

CRP (Citizen Readiness Program). This aspect warrants a paragraph in a foundational tactical skills manual. A Citizen Readiness Program is just the catch-all term for making sure you have all your administrative tasks taken care of ahead of time. Getting through the initial CRP (Citizen Readiness Program) quickly will put your mind at ease by knocking out administrative and medical readiness to allow you to focus on training. These initial admin tasks are not sexy by any means. They are tasks we tend to blow off but getting them squared away can alleviate a lot of the day to day stressors in life and allow you to focus on your preparation. The last thing a leader needs is a fighter with personal problems, the last thing your teammates need is a fireteam member who is combat ineffective because you have an infected tooth that could have taken care of before X Hour. Taking care of all those small tasks now will save you having to expend mental energy on them after you get the text message *"It's Happening"*. This CRP phase of the process is self-directed and typically includes medical preventative measures and admin items to ensure that you and your family are squared away. The last thing you should be worried about when called upon is some menial administrative action that could have been addressed weeks or months ago. The goal of the CRP is to help you minimize distractions caused by "nuisance tasks" that can take up much needed focus during a crisis. Examples of tasks to add to your personal CRP list are things like being up to date on legitimate immunizations (you know...real traditional ones like tetanus, polio etc). Having a will drawn up, consolidating critical information and documents, making backup encrypted hard drives, getting all dental issues fixed and up to date, new eyeglass prescriptions and getting two new pairs (if needed), etc etc. Bottom line is if you have a task that requires an appointment or entails filling out a form go do it this month and get it knocked out.

Acquire/Update Arms and Equipment

Equipping Yourself and Others. This section briefly addresses personal and team equipping principles; preparing your home or safe location is beyond the scope of this manual. Your clothing and equipment choice must be done through the lens of the limited logistics we discussed earlier and must be optimized for functionality, durability, and situational appropriateness (eg not wearing a large ALICE pack and Flecktarn parka while conducting a covert recon of a populated urban area). Parts commonality and gear similarity in your group will be critical in the anticipated scenario. It does not have to be lock-step in your group; your team doesn't need to look like a basic training formation with total uniformity, but common sense should overcome ego and an emotional attachment to a particular uniform, gear, or weapon setup. Blending in with the physical and human elements in your AO must be constantly assessed and adjusted for.

Camouflage. Depending on your mission, you may need to wear camouflage to blend in with your surroundings. Look for a camouflage pattern that matches the environment, such as a woodland or digital pattern appropriate to the terrain and season. Choose clothing that is made from durable fabric and has plenty of pockets for storage. Choosing an obscure camo pattern for you and/or your team or only having one pattern may be a mistake. Standing out or being known as the Swiss Alpenflage Gang is counter to sound tactical and OPSEC practices. Choosing a common pattern such as Multicam, M81, AOR2, or whatever is in wide use in your AO is a good starting point. Changing patterns as the seasons and background colors shift is always advised, do not handcuff yourself to a single camo pattern in your group. IFF (Identification, Friend or Foe) is also a real requirement. Fratricide (the killing or injuring of a friendly mistaken as an enemy) is inexcusable.

Things will change over time of course and adjustments will be required for uniform and IFF marking. We are just looking for a viable, reasonable starting point. Be sure to consider the IR properties of the clothing and gear as well, some manufacturers use dyes that will almost glow when observed through night vision devices. Used military camo is a great source, however prior care and washing practices may have damaged the IR compliant properties. Using detergents that are scent free and without UV brighteners is necessary (more on this later in the SLLS discussion). Many deer hunting detergents are good for tactical clothing, just double check to ensure they have no UV brighteners.

Be your own CIF. One of the resource constraints we have is there is no externally funded Central Issue Facility (CIF) for the Citizen. We are self-supporting and self-equipping; it is just the nature of the beast. Having sets (not just "a" set) of equipment will be of great value to you as you adapt to changing mission requirements. Chest rigs, plate carriers to include both slick and full up with load carriage, and durable packs are all required. The ability to swap to a light chest rig for a recon mission vs having to carry a full heavy rig is worth the added budgetary cost and storage requirements. There will also be members of your community that will be pressed into the fight without gear and weapons. While we may dislike the idea of taking care of someone who has not done their homework or has failed to commit resources to prepare, supplying them with a surplus LBE and a spare AR may be necessary. We are not saying expend resources during the preparation phase to account for someone else's failures, but we all have piles of gear that we can cobble together for a community member that has the potential to be a good fighter. Marking a storage bin as giveaway items is good practice and will be an asset for your neighborhood / community. Looking at it from another angle, we may even have to supply one of our own squared away teammates who has become separated from his or her gear enroute to your location when the bad event is on us.

Gear quality and budgeting for kit. We all need to have a method or decision criteria we use to determine if the *quality* as well as the design features of a particular item meets our needs. "Learned helplessness" has been a thing in the community, researching and applying critical thinking to your requirements (gear or otherwise) is the way forward. The best place to start is the question of meeting real, legitimate requirements. If an item does not meet your individual or group needs then it should immediately be dismissed as an option. This leaves us only with the selections that meet our requirements. Within these remaining options there are material selections that will be optimal, exceed, or are adequate / good enough. When we choose an item we always mentally trade off features, attributes, and cost. To arrive at the best solution available you should make that process a deliberate one with conscious thought. This is not a long and complicated approach, just an intentional one. *Determine what role the item has (the requirement) and what are the consequences if it fails.* Not all items are of equal importance for the Professional Citizen. When you take an objective look at your requirements the non-negotiable items are revealed by evaluating the consequences of failure for that item. The severity of failure will tell the actual importance of the item. For example, "If my chest rig falls apart and I lose my mags during a patrol it will cause mission failure" carries a lot more weight than "if my range bag strap breaks my buds will laugh at me while I'm policing all my stuff up". Mental wargames like this will assist with the decision process and uncover solutions to problems that may not be as important as we thought they were. ***Will the failure of the item result in severe injury or loss of life, mission failure, or significant damage to other equipment that cannot be easily replaced?*** If the answer is "yes" to

14

any of these questions, then the item is critical in nature and quality should not be compromised for cost. The lower limits of "good enough" become much higher for critical items, and they should not only be managed for quality they should become a priority for purchase. For non-critical items the criteria slips down to a "good enough" status, failure of these items will be an inconvenience but they are not mission critical. Just remember there still exists a floor where you cross over from "good enough" to single use junk. Good enough solutions are just that: good enough. Do your research, don't get sucked into the emotional defense of community brand loyalty without the facts. It might save you buying sub-standard solutions, but more importantly it will promote self-awareness, help define what is truly essential to your role, and free up budget resources to apply to the critical items that have little margin for error.

Individual Training

Individual Training is this manual's focus. There are obvious crossovers to collective (team, squad, platoon size) tasks as everything nests together. Individual tasks you acquire over time will feed into and support the larger collective tasks your team will conduct. To be a member of a good fire team you must all have a common or shared set of task proficiencies. Everyone must be able to shoot, move, communicate, land nav, treat a casualty etc. The foundations for these skills are based on military tasks, however we will not operate as a regular force so we must adapt these regular forces skills and tasks to our needs. Irregulars / guerillas supporting US forces and local authorities during emergencies will be expected to perform missions with limited or no resources; we will adapt our skillsets to operate in such an environment.

The remainder of the Road to War process builds on these initial skills and preparations. Other references in the

Professional Citizen series include collective and small unit training and tactics and more complex skills (eg patrolling, survival skills etc).

Be willing, and don't be afraid to train. Reading and consuming content is just the starting point, you must get outside and do the work. Some individual skills can be started alone, but **you must find training partners and local groups.** Checking egos and knowing that training in isolation is a blind spot for many in the community. You will be surprised at how willing and forgiving fellow Citizens can be. If they are unwilling to assist a good person of character and one with real desire to improve, they are probably not someone you would want to rely on in a crisis. Move on, there are plenty of others like you out in the community. The gatekeeping and "if you know, you know" type of individuals are plentiful (especially on socials and forums), but there are some really great folks in this community that are of the right mindset - and live the Professional Citizen life. But you have to put yourself out there and just get started.

The Framework

The Professional Citizen must be able to **Shoot, Move, Communicate,** and **Survive.** **Lead** is the fifth pillar; this is discussed in depth in the Citizen Leadership Manual. There are countless tasks and drills that reside inside these four main categories, far too many to train in one session, one week or one month. Individual proficiency in these tasks must be built systematically but also attacked with an extreme sense of urgency. Make no mistake - hard times are coming. Time is critical, every day used or wasted will make a difference.

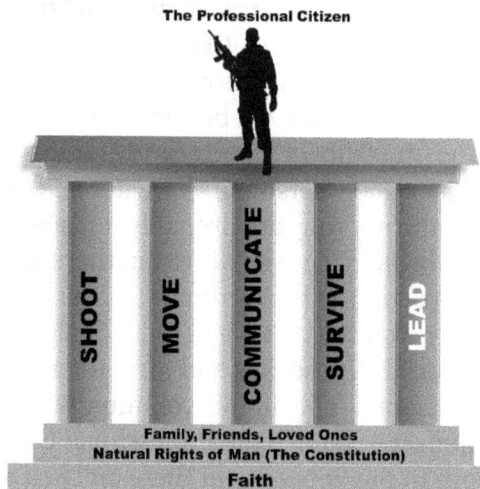

The Professional Citizen

SHOOT · MOVE · COMMUNICATE · SURVIVE · LEAD

Family, Friends, Loved Ones
Natural Rights of Man (The Constitution)
Faith

There will be no time to go back and train on a task during the days of mission execution after you receive the *"It's happening!"* text or message. It is easy to become overwhelmed by the sheer volume of this effort, taking it on with a deliberate approach is the optimal path to success. An approach to this is building a groundwork of basic skills that is constantly expanded by adding difficulty. You will do this through task complexity, combining tasks, collective training (team and unit) and training under adverse conditions such as bad weather and darkness. The Pro Citizen skillset is built on a holistic / whole person concept, an effective fighter is built on a solid footing of serving something larger than himself. This might come in the form of his faith, family, friends, and the pre-existing individual natural rights of man reflected in the US Constitution.

Shoot (Chapter 2)

We are a Nation of Riflemen. The musket that developed into the rifle that developed into the semi-auto self-loading carbine is the professional Citizen's weapon system. From the Kentucky Long Rifle to the modern AR/M4 series carbine our Citizens have relied on the rifle to harvest game, defend family and home, and to fight as part of a larger force. The ability to safely carry and expertly employ both the semi auto standard capacity pistol and the magazine fed rifle/carbine is one of the foundations of the Professional Citizen. This chapter will cover some of the foundational technical and tactical tasks associated with typical Citizen firearms selection, pre-combat actions, and basic employment.

Move (Chapter 3)

The Professional Citizen must be capable of moving quickly, quietly, and tactically over all navigable terrain. He must be capable of stealthy individual movement and have the skills to maneuver as a member of a team or unit. He must have the physical ability to carry a combat load (including loads with longer term sustainment) cross country while navigating day or night. He must be proficient in dead reckoning, terrain association, and route planning in a tactical environment and "read" terrain. We must anticipate the loss of navigation technology either by system failure or a conscious decision not to use it due to enemy cyber and geo locating capabilities. This chapter will discuss in detail individual dismounted movement, basic team formations, and foundational land navigation tasks.

Communicate (Chapter 4)

The Pro Citizen must be able to quickly and accurately convey combat information that will become actionable intelligence for themselves as well as small unit, regional, and community leadership. The Citizen must be able to communicate with visual signals, radio traffic, written messages and standard reports to provide timely information to others. This chapter is an overview of communication tenets and lays out visual and audible signals, reporting formats and procedures, radio procedures and standard brevity codes and words as well as radio communication basics.

Survive (Chapter 5)

The probability of limited or non-existent logistical capabilities is a reality we must be prepared for. Pro Citizens must self-sustain and have the skills to be a contributing member of their small unit's sustainment efforts. Medical skills that will allow the Pro Citizen to treat injury and prevent disease are paramount. You must be able to survive in austere field environments in severe weather conditions. In this chapter basic sustainment concepts are introduced; detailed procedures and skills are discussed in other manuals.

What Can we expect to be called upon to do during a security crisis / kinetic event?

Mission Profile. The Pro Citizen will be required to perform unpredictable and varied tasks to accomplish his mission and protect his loved ones. Odds are far greater that we will be pressed into service for some community crisis or neighbor event vs responding to an invasion or war on US soil. The non-kinetic events on this spectrum are a bit easier to prepare for and can effectively be addressed with a critically thinking human being who has done a little bit of prep work beforehand. While tactical requirements in a regional or at scale kinetic event are far less likely, the preparation requirements are far more complicated than being ready for a power outage. This complexity is why we must focus our efforts squarely on training for tactical missions. The larger categories of this tactical mission set will most likely include unconventional/guerilla versions of conducting Reconnaissance (Recon) operations, performing Security operations, conducting limited Attacks in the form of Ambushes or limited Raids, and Defending.

Of these four, the Reconnaissance and Security (R&S) will most likely be the backbone of the Citizen mission profile. The execution of R&S provides information, gains time, and can create space to adjust to changing situations. It allows us to react to opportunities and danger and enables leaders to transition to future operations. For example, Pro Citizens conducting reconnaissance will support leaders to transition from defense (security) to offensive operations. Pro Citizens organized in scout teams can provide the main force or homestead /neighborhood critical information and early warning of threat activities to afford them enough time to react and space to prepare or maneuver their forces into ambush sites and blocking positions. Citizens in the Scout role will enable leaders / decision makers to determine enemy intent, enemy's willingness to fight, and preserve freedom of maneuver for friendly forces.

During high threat conditions the Citizen may be required to conduct offensive operations, however the risk of becoming decisively engaged by the enemy increases - and so does the risk. The imperative is to remain elusive / undetected and not become decisively engaged unless there is an overwhelming possibility of success. Hit and run tactics, making widows, breaking the morale of threat forces, and then simply melting away into the terrain or community are the tactics of the irregular force. The Pro Citizen (even in larger teams) is neither equipped nor capable of sustained conventional combat operations against a larger / superior force. The cost of prolonged direct fire contact is too great for an irregular force to bear. Stealth, small unit and team movements, and an aversion to becoming decisively engaged are how we must view any future fight with a regular or organized force that threatens the US and our communities.

The unsupported and unfamiliar ecosystem

On the spectrum of the possible, a full-blown invasion or occupation of US soil by forces hostile to Citizens has a spot in the unimaginable category for most. However, it carries the most risk and dangerous consequences of any possible scenario for our society and way of life (arguably this is closer to being a probability than in any other time during our lives). A role inside of this will require checking normal mode bias at the door; the faster you recognize the change and accept the fact that life will never be the same again the better. There are imperatives that the externally unsupported fighter must wrap your head around.

Among these imperatives are:

This is not a game, it is not apocalypse porn. Beware potential team members who think otherwise and guard your own thoughts against these ideas. If and when this comes to our soil it will be horrible; this is not something we should want to happen or fantasize about.

You will fight with what you have right now. This includes hardware (gear, weapons) and software (the training and knowledge you have embedded prior to the event). Counting on belt-feds lying about and figuring out a comms plan in stride is wishful thinking.

You and your group / local network are your own logistics system. The support you receive will only be as good (or bad) as the means and methods that you put in place.

You will not fight "as if". You will not fight as if you are a conventional force with conventional combat multipliers or assets, nor will you use conventional tactics. History is loaded with examples of unconventional forces that attempted to adopt the tactics of their conventional foes. To put it bluntly, it did not go well. Embrace perceived shortfalls and make them strengths.

Variables of the Operational Environment

The operational environment evolves as each mission progresses. The two types of variables are operational and mission. Professional Citizens use **operational variables** to analyze and understand a specific operational environment and use **mission variables** to focus on specific elements during mission analysis. The operational variables are "big picture" elements and include political, military, economic, social, information, infrastructure, physical environment, and time. These will all influence and affect our operations, while worth mentioning for awareness these are outside the scope of this manual.

METT-TC. The second type of variables are *mission variables*, which describe characteristics of the Area of Operations (AO) focusing on how they affect a mission. The mission variables are **M**ission, **E**nemy, **T**errain and **W**eather, **T**roops and support available, **T**ime available, and **C**ivil considerations (METT-TC). We will briefly touch on METT-TC, application of this mnemonic is more complex but having a basic understanding is where we will start. METT-TC is used primarily as a framework to aid in analyzing a situation, prioritizing resources, and making informed decisions to accomplish your mission. The "**E**" in METT-TC is always the focus; orienting the solution on the threat or enemy must always be the main thing. Become an expert in quickly developing your assessments based on METT-TC variables. This cannot be stressed enough. METT-TC can be looked at as overused or a "catch all" that has been misused and over referenced by the tactical community. It is one of the best basic assessment and decision making frameworks available. It is easy to learn, is simple, and can be recalled when tired, hungry, and wet. It not only provides a framework for mission planning, it can serve to inform pre-execution preparations such as gear and body armor selection, weapon type, ammunition types, stockpile levels- the list goes on. Applied properly, it will serve you well. Troop Leading Procedures (TLPs) and the Military Decision Making Process (MDMP) are far more complex than just using METT-TC. These are discussed in depth in other Pro Citizen references.

Becoming proficient at applying METT-TC is the best way for a Citizen to start the journey of becoming a tactical decision making expert.

Mission. The first component, Mission, refers to the purpose of the operation (the main action you are taking during a given period of time) and the tasks that must be accomplished. Assessing the mission includes understanding the leader's intent (your leadership and one level above if applicable) and the desired end state.

Enemy. The second component or METT-TC, Enemy, involves analyzing the capabilities and intentions of opposing threats, including strength, tactics, and weaknesses. This assessment will drive everything we do. It must also account for forces or individuals who are neutral that may potentially become a threat or an actor inside our Area of Operations (AO).

Threats. A threat is any combination of actors, entities, or forces that have the capability and intent to harm U.S. Citizens, friendly US forces, Constitutional Interests, or the homeland. Threats may include individuals, groups of individuals, anti-US insurgents under the guise of protests, paramilitary or military forces, nation-states, or national alliances.

Regular Threats. These include peer and near-peer military threats that train, equip, and organize to conduct combined arms maneuver. They employ modern and modernized weapons systems capable of defeating or competing with similar U.S. systems.

Irregular Threats. Includes insurgent, guerilla, terrorist, paramilitary, and criminal organizations. They attempt to win by exhausting U.S. national and local political will by inflicting long-term expenditure of American lives   and influencing the social and political views of the people from within. A hybrid threat is one that employs a combination of regular and irregular forces.

Terrain and Weather. The third component of METT-TC is Terrain and Weather. This is the physical environment in which the operation will take place, including the impact of the environment on the mission, such as the effects of terrain or adverse weather conditions on mobility and visibility. Terrain and weather effects should be accounted for from both the friendly and threat perspective. Using advantages generated by weather and darkness can be used when planning missions or your static security posture. Terrain analysis is discussed later in the navigation section of Chapter 3 using the acronym OAKOC.

Troops and Support Available. The fourth component of METT-TC involves assessing the strengths and weaknesses of one's own forces (teammates, family members, supporting or supported local friendly forces), including manpower, their capabilities, readiness, equipment, and supplies. It includes the enablers that friendly US forces can supply. It is especially applicable to knowing yourself and assessing current capabilities. Lack of food, sleep deprivation, skills and abilities are all accounted for under this category. This is not just an acknowledgement; it must be applied to mission planning. Eg lightening of approach loads, allocating roles for a mission, sustainment requirements, accounting for sleep cycles, etc.

Time Available. The fifth component of METT-TC. This refers to the amount of time available to plan and execute the mission, including deadlines and the overall pace of operations. This component is critical in determining the feasibility of the mission and the resources required to accomplish it. Using the *one third, two thirds rule* is a good method to allocate planning time. This means you as a leader only consume 1/3 of the available planning and prep time while allocating 2/3 to your subordinates. 1/5, 4/5 is a better goal and will go a long way to best supporting subordinates. Parallel planning and pushing information down in the form of detailed Warning Orders (WARNOs) to subordinates will enable this greatly. Assess the time available for not only for planning and preparing, but also executing tasks and operations. This includes the time required to assemble, deploy, and maneuver in relationship to the enemy and conditions as well as available light data if the mission requires movement only during hours of darkness.

Civil Considerations. The final component of METT-TC is accounting for Civil considerations in your AO. In our environment this will be far more complex to account for than any prior military operation you may have participated in. Civil authorities (law enforcement, local political leaders, community / new tribal leaders), gang leaders and criminal enterprises will all impact our operations. You will have to account for covert and overt use of uniform and equipment during missions around the local population. We must be aware and tuned in to local population sentiment and attitudes toward us, local authorities, and each other. In-fighting and tribalism will be a thing, knowing who is who and considering the impact on our mission is essential. Having a mission disrupted or becoming compromised during reconnaissance because we did not understand the local population is unacceptable.

Gear and Equipment

Placing the gear and equipment selection section here may seem out of place, why wasn't it addressed under the Road to War / acquiring equipment section? It is placed here after METT-TC because of the requirements-based mindset we need to cultivate (and a paradigm we need to break in the community). Assessing your anticipated mission profile and addressing METT-TC and its application to your situation is preemptive to buying gear. We have all heard it: *"mission drives the gear"*; for some it is just a saying learned from the internet or a class without truly understanding what it means. The Pro Citizen must take it to heart and apply it. We anticipate that our mission set will be heavy on reconnaissance and security with minimal direct action and CQB. We will seek to avoid direct fire contact, not seek it out. We avoid it unless we are forced into it or initiate only when there are overwhelming odds that we will succeed. Yes, we must be prepared to fight and win under any conditions, but the approach will be a bit less conventional than many would visualize. Your gear and firearm choices must be in tune with this mission set to support your legitimate, real world anticipated requirements. Requirements that you have chosen based on research, lessons learned during training, personal and regional assessments, and wargaming.

Shakedown. Put your kit on and move cross country, practice and train on the range in your kit, and stay out in inclement weather with what you have in your setup. Gear setups that interfere with physical movement, create miserable hot spots, are ill-fitting or flops around should be discovered early on through use. What seems like a good kit idea may be abject misery for you when put into use. Drop leg loads (research shows that adding weight to your legs increases your energy expenditure by 4% per pound), rucks that rub or interfere with fighting load carriage, or boots that don't fit properly are disastrous when you can't come back home and swap them out. Accessing a canteen or water bottle without having to doff your entire kit, having ready access to maps and compass, or having your radio get turned on or off by pouch or gear interference are all things that should be discovered and remedied before X Hour.

Night Vision for Citizens. Image intensifying night vision (NV) systems are expensive, there is nothing cheap that is worthwhile in the night vision realm. But the capability far outweighs the cost. The advantages of NV during darkness are exponential, even when using "just" a PVS14. A quality PVS14 on a great mount is all that is truly required, duals are an enhanced capability but there is no comparison between the jump in capability from no NV to monocular and monocular to duals. The 14 has the advantage in cost and weight and is well suited for monocular (handheld) surveillance purposes.

The best advice we can offer is to work with a reputable dealer who builds their own units that you can call directly and have a conversation with. This will be doubly important if you go to a bridged unit later on as they will need to match your current PVS 14 to a second unit.

We recommend the lowest entry point to be a quality Gen 3 green phosphor PVS14. Digital night vision systems and past generation image intensifiers should not be considered for serious use.

White phosphor WP (not *phosphorous*) or green phosphor (GP) is a personal preference. White is a bit easier for long duration use, but a significant objective difference between the two colors has not been scientifically proven. A green tube with comparable specs will oftentimes be less expensive, so trading higher WP cost for an equal or higher performance GP unit is always a viable option.

Dual tubes are sexy and can be a super-flex on social media but they are incredibly expensive. It is better to have a high quality, clean PVS14 unit than a set of duals built on lower quality tubes that you purchased based on price while sacrificing quality. The quality difference between an $8k set and a $10-12k set of duals is substantial, and spending five figures on NV is a hard pill to swallow (pricing examples as of publication date, we may all laugh at the dollar amounts when we read this example in a few years). We won't dive too deep into NV selection, just be cautious about chasing specifications. A high FOM unit may be great, but there are other specs that may add up to make a lower FOM unit the better option. Educating yourself on how tube specifications become advantages or disadvantages as they are applied to your anticipated mission variables.

Use caution with selecting the housing if you choose the dual tube route, less expensive RNVG and BNVD housings do not allow for single tube articulation. Being able to rotate one of the two tubes out of your line of sight during missions is desirable. Older style housings will not allow you to do this, the only two modes with these housing types are both tubes flipped up or both flipped down.

Mounts. Quality mounts are ridiculously expensive, but they are a necessity. Surplus rhino mounts with bayonet couplers may work for a 14, but they are outdated and will not provide a stable mount for you.

Night vision technology is an investment in capability, just be sure to choose something that is requirements based. Image intensifiers, thermals, hybrid systems are all available on the open market and serve very distinct purposes. Don't overthink a purchase but certainly have a deliberate approach to making your choice. Night vision hardware and night operations are a complex subject that you will need to research thoroughly and train on extensively as you build your capabilities.

Helmet. Helmets can provide ballistic or impact (bump) protection and serve to mount hearing protection, communications, night vision, and IFF (Identify Friend or Foe) devices. There are two general categories with countless subsets of models, the two categories being ballistic and non-ballistic.

Ballistic helmets should be selected first and foremost from reputable US manufacturers. Most "NIJ tested" (not NIJ certified) offshore manufactured helmets should be avoided. They are indeed cheaper than a good quality US helmet and they will look the same /just as good, but they have poor quality control and are potentially dangerous. Choosing a ballistic over a bump helmet has a weight and cost penalty but will provide pistol (some models can provide limited rifle) and frag level protection. Ballistic helmets will be a bit hotter than a bump helmet but may require less counterweight than a bump helmet when wearing night vision systems. Bump helmets can be a good means for night vision mounting, just be cautious when choosing a brand and ensure the mounting interface is of high quality. You are placing an expensive night vision unit on what is essentially a lever that exerts torque and weight on the mount interface, you want to minimize the chances of failure or loss of your NV unit.

Load Carriage

Chest Rig or Plate Carrier. The answer to this is "yes". The Professional Citizen will need a chest rig / load bearing harness as well as a plate carrier (PC) to hold rifle rated body armor (hard plate armor). You can opt to add a battle belt / war belt to the chest rig or plate carrier for additional capability.

Chest Rig. A chest rig is a necessity for the Pro Citizen. The chest rig is not a do everything piece of kit, the chest rig allows us to carry magazines and a limited number of additional items. Handheld lights, radio, maps, night vision, IFAKs, admin items, small tools or knives, and maybe a handgun can all be configured on a good quality chest rig. Chest rigs are limited in their ability to carry sustainment items and provide no ballistic protection, but they are light and the wearer retains a lot of the mobility required by a fighter. You can opt to wear a battle belt to add to the overall capability when wearing a chest rig (recommend staying light as possible for belts without suspender systems). As with all things there are tradeoffs that we need to filter through our METT-TC analysis. We must plan to carry the bulkier and heavier fighting load related items in an assault pack when we choose the chest rig. *Micro rigs have their place on the range or directly mounted to a plate carrier, however a stand alone micro rig is not the best choice for our anticipated mission set due to its limitations.

A chest rig like this LSR is perfect for the Pro Citizen scout role. Mags, IFAK and a radio on board. You will have to carry sustainment and additional items in an assault pack, but the chest rig is the way to go for light and fast movement.

Chest Rig (CR)
Advantages
Good mobility
Light, Comfortable
Access to magazines
Disadvantages
Limits chest airflow in humid/hot weather
Pack required for sustainment, shelter, helmet
Application
Recon
When direct fire contact is possible but not likely
Vehicle Use

31

Load Bearing Equipment (LBE). The LBE is a proven design and can be easily sourced using high quality US Army surplus items or built from modern components. Modifications and creative configurations can provide outstanding results for Citizens. The LBE carries and distributes weight very well and interfaces with framed and frameless packs; they interface very well with framed Medium and Large ALICE rucks. Ideal for tropic conditions and hot weather, the LBE is comfortable and does not greatly inhibit mobility. The tradeoff is the bulk is all located in a ring around the user's waist so movement through narrow spaces or mounted operations can be less than optimal.

Load Bearing Equipment (LBE)
Advantages
Comfortable / distributes weight to hips
Light
Interfaces / worn easily with ruck
Disadvantages
Poor for mounted / vehicle use
Pack required for added sustainment, shelter, helmet
Application
Recon
Direct fire contact possible but is not likely
Jungle /tropical areas
Training Range use

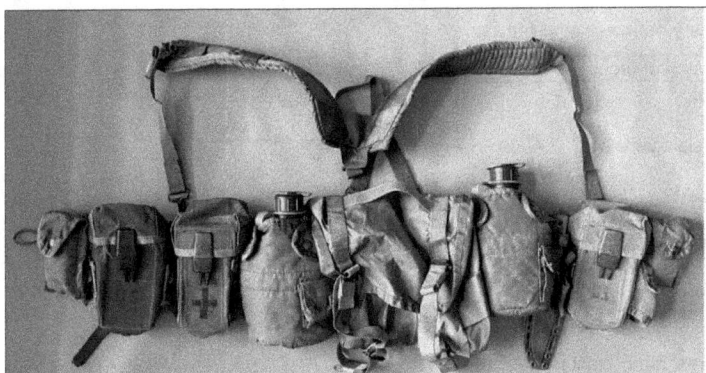

The respected LBE. The ALICE system LBE is still a great belt and suspender systems (if assembled and used properly). Durable, cool in hot weather, and readily available for purchase or trade. There are drawbacks of course, but if adjusted and setup properly they can still make outstanding rigs. Updated pouches, belt sleeves etc can be added to update the setup as desired. Even if you don't use yours anymore store it away in the contingency handout bin. Someone may be glad to have it one day.

Plate Carrier (PC) and hard plate armor. The plate carrier is the newest option among the four and is the only system that can provide ballistic protection. Plate carriers have been around for decades and have morphed from simple two-piece designs that just carried the armor plates to full featured pieces of nylon kit with PALS webbing for days. They can be worn slick by themselves, with chest rigs or LBE systems over top, or directly mount a full suite of pouches and load carrying items via the MOLLE system. Do not wear a plate or armor carrier without armor; they make terrible load carriage platforms when mis-used in this manner. The plate or armor/plate carrier systems have the distinct advantage of ballistic protection in that they allow us to carry a small piece of our own mobile cover. This capability comes at a cost, both financial and weight.

Plate Carrier (PC)
Advantages
Ballistic Protection
Good Mobility
Weight Distribution
Use in vehicles / mounted
Disadvantages
Hot
Weight penalty
Cost
Application
Urban areas
Direct action
Direct fire contact is likely
In Vehicles

The BC19 is a good example of a full featured modern plate carrier. Fighting load can be direct attached or worn over top of a slick carrier. Either way is good to go.

Armor is an absolute necessity for a Pro Citizen; however the art and science is knowing when and where that necessity will arise so you can tailor your loadout based on mission requirements.

Ballistic plates have significantly decreased in weight over the years as materials have become lighter, but there is still a weight penalty to carry armor. There are several options that are available, do some homework and select based on your requirements.

Selecting Armor. Levels of armor protection. Until the NIJ updates their standards (as of this publication NIJ has still not adopted the new proposed standard) we are stuck with an outdated rating system and need to be smart about how we select our armor. Under current standards Level II and IIIA are rated to stop specific pistol rounds, Level III and Level IV are rifle certification ratings. Ratings are not progressive (eg a Level IV plate is not automatically assumed to be Level III certified). Soft armor stops pistol, with current materials available it takes hard plate to stop rifle rounds.

NIJ Certified rifle rated plates stop 6 hits of 7.62 ball (Level III) or 1 hit of .30cal Armor Piercing (Level IV). Specifically:

Level III - Six (6) 7.62x51 NATO ball (147gr) hits (so essentially all Level III plates are "multi-hit") . Fired from 15m at 2,780fps.

Level IV - certified one (yes ONE) hit of 30cal M2 Armor Piercing (30.06, 166gr) fired from 15m at 2,880fps. A Level IV certification does not ensure the plate will withstand Level III as well the Level IV threat (it is not a progressive rating). However, this does not indicate that the plate will fail against a Level III threat, it simply indicates that the plate has not been tested and certified against it.

SAPI - US gov't rating, 3 rounds of ball, 3 rounds of AP, exact test conditions are not available to the public. Not NIJ certified.

Special Threat / non NIJ - there are some great special threat plates out there that are not NIJ certified (remember the only certifications are for M80 ball and M2 AP). Reputable manufacturers have spent a lot of time and money to develop some innovative solutions that may not be NIJ certified but may fit your requirements better than an NIJ certified plate.

Level III+, IV+ etc. Some manufacturers use these ratings to describe capabilities of their plates, but there is no such thing as a "+" rating. Usually used when describing a plate material that is known to have issues with a particular threat (eg M855 and poly) that the manufacturer has modified with an additional material to mitigate the non-NIJ threat shortfall. There are some reputable manufacturers that use this rating, but it is for marketing purposes only and will most likely go away when NIJ finally adopts the updated standards.

Part of selecting armor is understanding materials and the associated benefits and downsides. We typically select ceramic or a hybrid construction (ceramic and polyethylene (PE)) for hard plate and avoid steel. Steel is cheap, durable, and available however the market is flooded with steel products that do not meet our requirements.

The "armor triangle" is the relationship between price, weight, and performance. When balancing *Performance, Weight, and Cost* for armor selection you can only choose two. Some materials are cheap but they tend to be heavy and can have performance issues. Lightweight, high performing plates are going to cost more due to the type of materials. Steel is the heaviest of all the materials, is vulnerable to high velocity FMJ rounds (eg XM193) under certain conditions, and anyone in the vicinity of a steel plate that is hit (to include the person wearing it) may be subject to ricochet. Compressed PE (poly) plates are very lightweight and some are even buoyant, but M855 will fully penetrate them. Pure PE has its place, but they are not recommended for our purposes. A high quality ceramic or ceramic hybrid armor plate is the best option for the Pro Citizen. The same rules for helmet selection apply to torso armor; buy from a reputable US based company, understand your requirements before you select a model, and avoid making your selection solely based on cost.

Shooter Cut (front in photo) and SAPI Cut (the rear plate in the photo) are personal preference. Some prefer shooter cut front, however using two SAPI cut plates is the most common. Range and training time wearing kit will alleviate potential issues.

Load Bearing Vest (LBV). The LBV was somewhat popular in the 1980's and 90's however it still can be a viable option for the Pro Citizen. There were several commercially built types that were more vest-like, but when we talk about LBV it is the short-lived but recognizable LBV88 that the Army developed. Any of the LBV and LBV-like setups share similar positive and negative characteristics. Among the downsides is the limited airflow and breathability of the vest. While there is a weight savings over a PC, the loss of ballistic protection with many of the same benefits it puts the LBV toward the back of the recommendation line until there are viable updated versions available. With the addition of a butt pack it becomes a 24-48 hour stand alone rig (METT-TC).

Load Bearing Vest (LBV)
Advantages
Good Weight Distribution
Can carry 24-48 hours of sustainment and expedient shelter
Places mags on chest with main load on waist
Disadvantages
Hot
Heavy
Does not interface well with ruck
Application
Recon
Direct fire contact possible but is not likely
Stand alone use for short duration missions

The LBV is still a viable option for some applications as long the user understands the limitations. (the belt is unattached and reversed for photo)

Load Definitions

Fighting load consists of the equipment (weapon, ammunition, helmet, body armor, water, etc.) and uniform (boots, clothing, etc) that is worn or carried directly on your person while maneuvering and fighting. A fighting load is only what you carry once contact has been made with the enemy. It consists of the essential items needed to accomplish your task during the engagement. Excessive approach loads of assaulting troops must be configured so the excess can be redistributed or shed (leaving only the fighting load) before or upon contact with the enemy. There is no way to predict this with 100 percent accuracy, but the more homework (recon, intel analysis, "battlefield calculus", and METT-TC analysis) we do prior to stepping off on a mission the more tailored (and potentially lighter) our loads will be.

Approach load consists of the fighting load plus a rucksack carried during a march or movement, which would contain additional water, ammunition, food, and other supplies for the duration of the mission. An approach load contains the operational essential equipment carried in addition to the fighting load. These items are dropped in an assault position, Objective Rally Point (ORP), or other rally point, before or upon contact with the enemy. On extended operations we must carry enough equipment and munitions to fight and exist until a planned resupply can take place. The approach load and fighting load will morph and blend together by mission. Be cognizant of the fact that you may carry the approach load almost the entire time during some reconnaissance missions.

Emergency Approach March Load. Circumstances could require us to carry excess loads (greater than 45 percent of body weight) such as approach marches through terrain impassable by vehicles or when ground transportation resources are not available or tactically sound. Although loads of up to 70 percent or more of an individual's body weight are feasible and can be carried by the well-conditioned, the bearer will quickly become fatigued or possibly injured. If possible, contact with the enemy should be avoided since march speeds will be slowed tremendously.

Loadout Weight. The only way to get a feel for the true burden of the loadout weight is to carry it over varied terrain for extended periods of time in hot, cold, and inclement weather. Heavy loads can have the opposite intended effect, they can severely reduce and not enhance combat effectiveness. Through our research and personal experience, we know that overloading individuals is not just an Army thing. Citizens have an even greater propensity for carrying way too much stuff. We have all heard "you can never carry too much ammo!" Avoid advice that makes such declarative statements; besides being cringy it ignores the fact that weight you can carry is finite and must be tailored to your mission. The primary consideration is not how much you can man up and physically carry, the professional move is to do your homework and determine how much you can carry without impaired combat effectiveness—either mentally or physically. It is a balancing act of what you need to prosecute the mission versus the physical detriment of that load. We know through common sense and experience that slower movement under contact will result in a much higher probability of being hit. The lighter you are, the faster you will be and the less likely you will be hit. Studies have shown every added pound of kit your sprint speed decreases about a half a percent (between .36 and .64 percent) *Paying For Weight In Blood: An Analysis of Weight and Protection Level of a Combat Load During Tactical Operations, Capt. Courtney Thompson 2019.* This means that for every 10 lbs you are 5 percent slower during a 3-5 second rush under fire. Under a 50 lb load (remember that weight is probably just a fighting load and includes boots, helmet etc) you are sprinting 25 percent slower than normal...while an enemy is shooting at you. Let that sink in as you work on your loadouts.

Travel light, freeze at night. *
SSG C. Moles

* let this serve you as a guide *and* a warning

As a general guideline, the weight of a fighting load should not exceed one-third of your body weight. This is known as the "33% rule" and the full Approach Load should remain under the 40% range. In some cases, you may need to carry more weight, but it is important to ensure that the load is distributed properly within your pack and load carrying equipment.

Yeah, we know. We have done it as well; carrying 130 lb (and then some) loads on our 165 lb body. It has been accepted practice to carry extreme weights on missions, overloading ourselves and our subordinates to the point of injury. These are the types of injuries that can be immediate and can render you or a team member combat ineffective. Fatigue from carrying excessive weight will decrease the performance of anyone no matter how physically fit, much less one who is operating remotely with little to no support outside of his own team. Being in an austere environment without support there is an inclination to carry everything you can, but doing so will break a man's body down quickly. The potential injuries and exhaustion of carrying heavy are precisely why we need to lighten the loads. According to an Army study from 2008 a third of medical evacuations from the field in OIF and OEF from 2004 to 2007 were for spinal, connective tissue, or musculoskeletal injuries; two times those from combat injuries. With the potentially limited medical evacuation systems in a post-event environment, we need to alleviate these injuries. Personnel quantity and quality will be our number one challenge, we must keep as many quality fighters in the fight as we can. Losing a team member from a patrol because he blew his knee out carrying an excessive load has the same short term operational effect as a WIA. Being deliberate and judicious about what we carry (or allow our troopers to carry) leaders will keep more fighters in the fight long term.

Army doctrine specifies a 48 pound fighting load and a 72 pound approach march load but these are violated willfully and regularly. We have paid the price for doing so through injuries and the resulting reduced combat power. A 2017 Government Accountability Office report reported Marine loads of up to 159 pounds, with an average of 117 pounds, and Army loads of up to 140 pounds, with an average of 119 pounds. Remember these are loads supporting troops that have full logistical systems and virtually endless access to combat multipliers such as artillery, mortars, and air power. An irregular force will have none of these assets other than what friendly US or your local organized force systems can provide. The limited firepower in a dismounted team of irregulars may drive our desire to carry even more than an organized force does, but doing so will create problems in the form of these injuries and decreased combat effectiveness through fatigue and slower movement.

Quick Tip: MRE's are not always the best solution for patrols. If you do choose commercial MRE's be sure to break them down into components during prep for combat to minimize weight, bulk, and trash.

Chapter 2
Shoot

Being both proficient *and* safe with your firearms is imperative. Team members must rely on you to dependably place accurate, controlled fire on the threat. They must know you are able to safely move in a tactical formation without muzzle sweeping their body. Your teammates must trust that you will not negligently discharge your firearm. Practice and train these skills to so they are done without conscious thought. Safe handling is not *the* task, it is just something that you do constantly while you are performing your job.

We will not address in depth manual of arms or a lot of firearm specific technical skills as the types of firearms are far too numerous to discuss in this manual. Overarching concepts in this chapter will apply no matter which specific firearms you choose.

Firearms safety procedures are sometimes scoffed at as "Fudd like" or having a flat range mentality. Quite the opposite is true if applied in context with a realistic mindset. To inadvertently shoot a family member, a teammate, or destroy a critical piece of equipment is a needless loss. Giving away your position with a negligent discharge can cost the lives of your team. A thinking, disciplined individual is the primary safety mechanism for all weapons under his control. There will be times when some rules *are* violated in tactical environments, we always want to ensure it is a conscious decision when doing so.

The firearm safety rules apply for every type of weapon system - and they apply when pulling security or guard duty, on the administrative practice range, in your house, or during a recon patrol in combat. The Citizen must maintain situational awareness of friendly forces to prevent fratricide; constantly assessing the environment to properly handle the firearm is the mark of a professional.

The Five Firearm Safety Rules (yes, we use 5)

0. Rule Zero. Be familiar with the mechanical functioning of the firearm you are about to handle.

Ego has consequences. If you are unfamiliar with a particular weapon or family of weapons you have no business carrying, clearing, manipulating, or handling it. Period. *More than a few negligent discharges have resulted from ego and ignorant handling* of striker fired pistols, 1911s, belt fed open bolt machine guns – the list goes on. There is no shame in admitting you are inexperienced with or do not know the components and functions of a particular firearm. Leave it alone and defer to someone who knows the mechanics of that firearm model.

1. Control your muzzle (Muzzle Awareness)

Avoid pointing your gun at anything you do not intend to destroy. In the event of a negligent discharge, no injury can occur if the muzzle is oriented in a safe direction. Know exactly where your muzzle is pointing (Muzzle Awareness), and always be in control of it. Ensure the path between the muzzle and target is clear of friendly forces, noncombatants, or anything else you do not want to shoot. When breaking this rule becomes situationally unavoidable (and there will be times when this happens), you must minimize the amount of time the muzzle is oriented toward people or objects you do not intend to shoot. This is 100 percent your responsibility, and only you can control it.

2. Treat All Guns as if they are loaded

Whenever you handle a firearm, or hand it to someone, always engage the mechanical safety, remove the source of ammo (magazine), open the action, and visually check the chamber to be certain they do not contain any ammunition if you know how to do so (see Rule Zero). Never assume a gun is unloaded when handed to you — check for yourself even if you just watched someone else clear it.

3. Keep your finger off the trigger until you have made the decision to shoot

Never touch the trigger on a firearm until you intend to shoot and keep your fingers away from the trigger while loading or unloading. Mechanical safety devices are not available on all types of weapons, but when they are present, do not solely rely upon them for safe operation knowing that mechanical safeties may fail. Never pull the trigger on any firearm with the safety on the "safe" position unless doing maintenance / performing a functions check on a cleared gun.

4. Always Be Sure of Your Target and What's Beyond It

Don't break the shot or fire the burst unless you know exactly where your rounds are going to hit. You must positively identify the target and know what is in front of and what is beyond it. We do this to prevent fratricide by ensuring rounds will not injure any friendly between, beyond, or near your intended target. This applies every single time and is especially important when shooting in support of / near a friendly maneuvering force. Target ID both day and night is critical to prevent fratricide (shooting or killing a friendly). There will be times when we will violate this rule as well, situation will dictate. While we will never blindly fire into an area there will be a need to fire into last known locations where an enemy was. For example, you engaged a threat near cover or concealment and are now unable to see him. You are reasonably sure he is still combat effective and is in the same general area where last engaged. The team needs to suppress or attempt to stop the threat even though you cannot positively id a target, so rule 4 gets conditionally suspended as we apply METT-TC and the current Weapons Control Status (WCS) that we will discuss later.

The AR pattern 5.56 is the recommended primary firearm for the Pro Citizen. They are common, lightweight, and low recoil. Ammo weight is more than manageable for a fighting load.

Selecting and Setting Up Your Guns. This can be a contentious subject, however there are principles that will assist with getting as close to an optimal solution as possible. If X Hour comes to pass there will most certainly be adverse conditions and limited availability of firearm related resources that you must account for.

Standardization

While this is an individual skills manual, setting conditions to be a functioning member or leader of a team must be accounted for in your initial training and preparation. This includes your weapon/ammo selection. Choosing survival and prepper type firearms requirements (.22 LR for example) is not within the scope of this manual, this section addresses specifically those firearms that will be needed for combat. The availability of ammo, firearms components and repair parts may be in short supply or non-existent during a large scale event. We have all seen the ammo and firearm shortages under somewhat normal conditions, these scarcities will be significantly worse as production and supply chains are disrupted. Selecting a common caliber for primary and secondary firearms will be critical as will choosing something that will probably be available in large quantities. If most of your local team/group trains and shoots 5.56 and 9mm then it would be wise to follow suit. Agreeing upon this now will go a long way toward mitigating problems that can arise after X Hour. While standardization is required, deleting variations of weapons and calibers that may cause an associated loss of capability should be carefully assessed. For example, a 7.62x51 DMR may be an outstanding combat multiplier in your team. You and your group may decide it is reasonable to add these even at the expense of logistically supporting a second rifle cartridge. The

Two Caliber Mix
7.62x39
7.62x54R

Two Caliber Mix
5.56
7.62x51

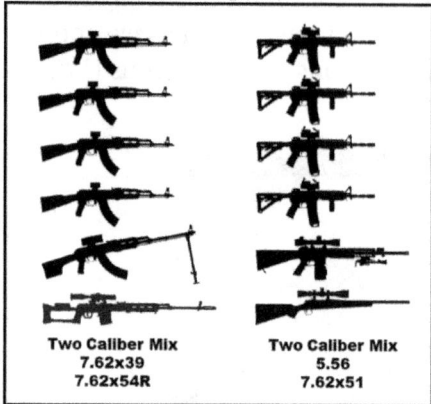

Weapon Mix Combo
Examples (Illustration Only)

diagram above is just an example of common cartridges and how you might choose to mix or not mix. At a minimum your family should be standardized and centered on one rifle caliber and one pistol caliber (caliber is not an encapsulating term, we just use an inaccurate shorthand term to denote specific cartridge selection). There is little to no increased capability in having different calibers for secondary weapons (pistols) in a team. Standardizing 9mm as the secondary weapons for your family and MAG is the logical choice.

Weapon Type Commonality. To be an effective force community groups must focus on having as much commonality as possible across like weapons. Having a "one off" gun in your family or group that would otherwise be caliber compatible with others in the team is not optimal (eg your stubborn uncle with a Mini 14). For same caliber long guns of different manufacture or model the magazines should be interchangeable. For example, mixing FALs and AR-10s as enhanced battle rifles is a sub-optimal solution: same ammo, different mags is a non-starter. A similar issue would be a mag fed 5.56 firearm that will not accept the entirety of STANAG / NATO or commercial magazine brands. Specific weapon choice aside, just bringing these mixed magazine types together (the Mini 14 / AR example) for your team would not be a good course of action.

The Gun. We as a community are great at expending our resources on guns. Dudes will buy random, obscure, and impractical firearms with regularity. Acquiring a niche or novelty firearm is best left to the collectors. We do not have the luxury of optimizing and chasing capabilities to obtain a purpose-built gun for *every* situation.

A high quality AR/M4 pattern 5.56 rifle or carbine should be the primary gun for every Professional Citizen.

Doing METT-TC analysis and wargaming with your family or team will uncover a discrete set of probabilities you may face. Outside of the requirement to have a specific covert or concealable long gun solution, optimizing a gun and optic for a single use case can be a trap we fall in to when we overthink the problem set. Having a 10.5 inch CQB carbine for room clearing, a 20 inch Designated Marksman Rifle with a 2-10x optic for long range shooting, and a 16 inch general purpose carbine for standard use is counterproductive. Not only will pursuing niche solutions consume resources, the odds of being able to choose which one is available when you need it for a specific situation are low. Choose quality, choose the best all-around solution for your requirements in your Area of Operation (AO), and most importantly learn and become an expert with the gun you have. Selecting the right firearm and accessories is a small portion of being capable and lethal; attempting to buy capability is a false solution. The combat effective individual is not the one with a basement full of ammo and a collection of dozens of boutique firearms. The Citizen who knows their rifle inside and out, has trained with it, has a current zero and recorded holdovers is who we want in our community and in our teams.

> "Beware the man with one gun. He can probably use it."
> — Jeff Cooper

Start with a high-quality trusted brand and train with it under field conditions. This necessity also applies to the accessories you choose. Most budgets are limited, waiting a couple of months to save and purchase a higher quality solution is always the better course of action. Meet *your* informed, researched requirements - not someone else's.

Cloning a system just because you like the way it looks or duplicating a particular setup because some "high-speed" unit or social media influencer used it is ill advised (we all know someone who has done so). Do not copy or clone someone else's setup just because it works for them.

Setting up your long gun

The rifle or carbine must be expertly built, either by the factory or by building one from a high quality parts kit if the builder is competent. A simple, light (as practically possible) high quality gun is the requirement. It should not be so lightweight that it sacrifices rate of fire / durability to shed weight. Pencil barrels and competition setups are not appropriate for tactical applications. At a bare minimum the Citizen rifle/carbine requires the addition of three enhancements to the quality base gun:

Sling. A high quality, US made two point adjustable sling is required for positive control of the long gun during physical activity, positional shooting, and used for retention when using both hands for other tasks. Attachment points along the gun are mostly personal preference, only time and repetition will tell which are best for you. Make sure you can tighten it down on your body and the sling placement does not interfere with reloads / manual of arms. On our team (to a man) we attach 9 oclock on the front and direct attach (thread through) to the outboard side of the stock at the rear. One and three point slings are not optimal, thankfully these have been relegated to the community memory hole for the most part.

Light. A weapon mounted light that is durable, waterproof, and can throw enough light to give the shooter a potential advantage over an adversary is the second required accessory. Choose a light that is typically used for rifle applications, nuances like spill and throw between good brands is mildly interesting but not critical for our application. Choose what meets your requirements. Using a pistol mounted light (Surefire U Series for example) is not the best solution for a carbine or rifle. They are great lights and are the perfect choice for defensive handguns, but a mounted rifle or scout type light is the best way to go.

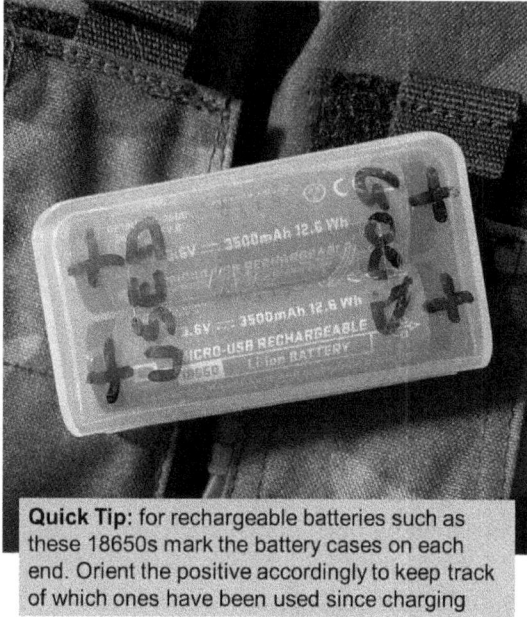

Quick Tip: for rechargeable batteries such as these 18650s mark the battery cases on each end. Orient the positive accordingly to keep track of which ones have been used since charging

Sight. You can use variable power optics, a magnifier, or fixed power optics. All have their advantages and disadvantages, just apply your terrain and anticipated mission to determine which one will serve you best then go train and practice with it. An electronic sight (Red Dot Sight – RDS, Holographic Weapon Sight -HWS) or a magnified optic such as a fixed power prism, ACOG or a variable power optic is highly recommended. Yes, iron sights are proven, rugged, require no power, and there are many shooters who can use irons effectively. However, the significant advantages gained by using modern technology are more than worth the monetary and weight cost.

Magnified optics provide a marked advantage; having some type of magnification will not make you shoot better, it will help you *see better*. Threats will be in shadows, in concealment possibly hundreds of meters away, and will do their best not to replicate the target-like qualities we are used to shooting at on the flat range. Being able to see into the shadows, get a magnified view of suspected positions, and spotting / identifying threats in concealment will all call for some form of magnification. Ensure you have a solution that will allow use of passive night vision as well. LPVOs and fixed power magnified optics do not typically work well with NVGs; plan on adding a micro RDS either offset or top mounted for passive NV shooting.

Signature and Camouflage

Muzzle devices. Flash hiders, compensators, and muzzle brakes (aka Blue Falcon Devices) are all types of muzzle devices. A flash hider and/or a suppressor is a requirement for any fighting rifle. Reducing the flash and sound as much as possible will prolong the time it takes an adversary to locate your position, especially so during hours of darkness. Muzzle brakes have no place on a fighting long gun. Even if used as a base host for a suppressor these become a liability with the inevitable removal of the suppressor. While brakes and comps can reduce recoil and split times on the competition range they are a liability to an individual or your team in combat. The increased signature and concussion imparted on teammates is not worth individual recoil mitigation. And firing a long gun in an enclosed space such as a vehicle or a room with a brake is a non-starter, even with ear pro.

Sound. Other than the obvious sound of firing there are other noise considerations when setting up a long gun. The materials and components on a firearm can be noisy when they rub or bang together in the woods, especially at night.

Attachment points and slings can be particularly noisy; sling adjustment hardware hitting a handguard or stock must be quieted with the use of 100 mph tape and/or 550 cord. Quality QD attachments are convenient and a solid way to attach your sling, but they are loud as the parts can rattle during movement; and even the good ones have been known to fail during missions. Using non QD methods to attach a sling are the preferred technique for a Citizen long gun, don't be afraid to use properly secured 550 cord to attach your sling. It is cheap, quiet, and absolutely field repairable.

Quick Tip: Sling attachment methods. Make sure yours is quiet and secure. 9 o'clock with 550 cord and direct attach to the outboard rear of the stock is our preferred method. Silent and secure.

Visual Signature. Painting your long gun(s) is prudent, a black carbine really stands out even under passive night vision. Ensure your individual, team, and group SOPs require painting all long guns. Excuses for not painting a carbine that may be pressed into combat are not valid. The "resale value" is a sad excuse for resisting the requirement, as is the marriage to the aesthetic finish of your guns. Any platform worthy of our purposes is gonna' look exactly like every other one when it rolls off the factory floor anyway.

There is no need to attach sniper-level scrim and camo material, but painting and disrupting the outline of your gun is a best practice. Flat paint with colorways that work for your AO is all that you need. Some paints may appear bright under passive NV, just do some personal testing and see what works and what doesn't under NV in the field. When adding camo material or texture keep it limited to ensure it does not/will not interfere with weapon functionality. We can also add small elements of texture (limited burlap or weapon tape) to further break up the silhouette, just don't overdo it. We do not want or need the moving tree look, just enough to diminish the gun's visual signature in your environment.

Tasks (procedures will differ for various types of primary weapons)

Unload and Clear. (verify with your specific model's manual), for an AR/M4 pattern rifle it will entail the following while keeping your finger off the trigger and muzzle pointed in a safe direction:

     - Attempt to place it on safe

     - If it will not go on safe move to next step

     - Remove the magazine

     - Lock the bolt or action to the rear

     - Place the gun on safe

     - Inspect the chamber visually and physically

     - If verified empty then release the bolt (control the charging handle forward is preferred)

Function Check. A function check should be performed each time the rifle is reassembled or any time there is a doubt as to whether the selector lever and rifle are functioning properly.

Clear (if it was not cleared immediately prior as part of this process)

Check Safe Function. Pull the charging handle to the rear and release. Place the selector on SAFE. Pull the trigger, and the hammer should not fall.

Check SEMI or FIRE. Place the selector on SEMI or FIRE. Pull the trigger and hold to the rear. The hammer should fall. Continue to hold the trigger to the rear, pull the charging handle to the rear, and release it, Release the trigger with a slow, smooth motion until the trigger is fully forward. The hammer should not fall and you will hear an audible "thunk" as it resets. Pull the trigger again. The hammer should fall.

Weapon Check. The weapon check is a visual inspection of the weapon and starts where the function check leaves off. The weapon check can be incorporated as part of your Pre Combat Checks.

Clear (if it was not cleared immediately prior as part of this process)

Verify it is the correct weapon (eg that it is yours and that you actually have the gun that you think you have). Sounds crazy right? Cold, tired, and hungry does some strange things to a man's awareness. Grabbing someone else's gun is a possibility especially if your team uses similar platforms.

Check attachment points of all aiming devices, equipment, and accessories by physical touch and witness marks (paint marks) to ensure everything is tight. Check the sling to ensure it is attached and serviceable. Check the adjustable stock is locked into the proper notch for you.

Check all optics and lasers switch status.

Check and reapply lubrication to the bolt carrier group (BCG) as necessary. Note that a weapon check is not detailed maintenance, just a final look before loading and conducting training or operations.

SLLS check. Check optics and light covers to ensure they are in place, check camouflage (paint or camo tape wrap). Sling the weapon over your chest in the low ready and shake/rotate by the pistol grip to verify it is as quiet as possible.

Fitment / interference / snag check between the gun, the load bearing equipment, and your ability to move between positions. Verify the proper adjustment of the sling; run it out to max length and make sure you can mount (shoulder) it and conduct reloads in various shooting positions. Then run the sling all the way back in to verify it secures tightly enough to your gear/torso.

Load. Starting with a cleared weapon IAW the procedure above, initial loading of the carbine during prep for combat, administrative, or range loading (eg you have time and are in a secure environment) entails the following while keeping your finger off the trigger and muzzle pointed in a safe direction:

Lock the bolt to the rear

Visually check the chamber (one more time)

Grip and insert the magazine, pushing (not slapping) it in until the magazine catch engages and holds the mag, then pull down to ensure it has locked in. *This "push/pull" is critical every time you insert a new magazine. It is easy for it not to fully catch and still (temporarily) stay in the mag well only to fall out when you fire the first round.*

Release the bolt catch (do not ride the charging handle, just use the bolt catch and let it go). This will increase the chances the bolt seats on the round in the chamber.

There are two ways to check for proper loading on an AR pattern gun. The first is a crossover check (M4 mags are double stack and feed from alternate sides). Remove the magazine and if you see (or feel at night) that the top round has changed sides in the mag the original top round is in the chamber. Ensure you use the push/pull method discussed before when you re-insert the mag; inserting a loaded mag on a closed bolt is known to interfere with proper seating. The second method is a chamber check by partially retracting the charging handle to see (or feel at night) that there is a round in the chamber. In this method releasing the charging handle may not completely reseat the round, an additional tap or two on the forward assist will ensure it rechambers fully and is ready to fire.

Zero

Refer to your technical manual for your firearm and optic(s) for specific technical zeroing procedures. For shooting positions and basic marksmanship tenets refer to any of the widely available manuals and publications available. While zeroing procedures vary across weapon models and optics there are some principles for zeroing a rifle or carbine that we must follow:

Understand the process. Know the "why" and "how". A zero is simply the intersection of your line of sight viewed through the sight(s) and the trajectory of the bullet. The point of aim and point of impact crossover only occurs twice in the bullet's path; once when the trajectory crosses the line of sight at the zero target and again at the far zero distance (number of meters or yards for your zero).

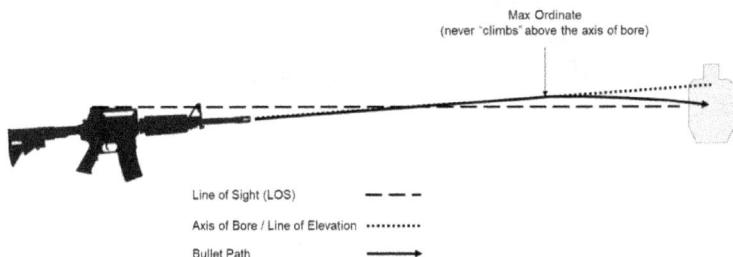

Max Ordinate
(never "climbs" above the axis of bore)

Line of Sight (LOS)   — — —

Axis of Bore / Line of Elevation   ··········

Bullet Path   ⟶

Bullets do not "climb"; once exiting the muzzle they are negatively affected by gravity and immediately begin to be pulled back down to earth. At first look it may seem like the trajectory climbs, but this is simply the fact that your barrel is canted upwards slightly in relation to your line of sight. This allows the bullet to arc in a manner that we can maximize to achieve more range than if the barrel and sights were aligned in parallel. Variables for zero include the ammunition, barrel length, height over bore of your sighting system (risers), and barometric pressure just to name a few. The zeroing process ensures you, your weapon, optic, and chosen ammunition are performing as expected at a specific range to target with the least amount of induced errors. Zeroing your firearm(s) is not a tactical task, this is a preparatory or maintenance task that needs to be managed and tracked.

Take your time. Rushing through the zero process is doing yourself a disservice. Respect the zero and set aside enough time to do so. Avoid the temptation of getting to the range a half hour before your shooting partners with an un zeroed gun; zeroing is a deliberate process that requires your undivided attention.

Be comfortable. For an initial zero, especially with a new firearm or optic ensure you have a stable shooting platform (either prone supported on a good quality shooting mat or from a bench with sandbags). You can refine your zero from other unsupported positions later if you think it will make a significant difference, but getting a good tight zero is the first goal.

Zero with your wartime ammo. You will most likely have multiple brands and types of ammo, ensure you have zero data and holdovers recorded for each. At a minimum you should zero with your typical training ammo, with XM193, and your warstock ammo (pick one war stock and stick to it for stockpiling, don't try to chase performance with this). XM193 is in common and widespread use, while this is not an optimal post-X Hour combat round, odds are it will be widely available, and you should have data for it with your gun. Zeroing for other rounds is encouraged, if nothing else so you understand how your rifle performs with them. Start the process with your typical training ammo and shoot/train for a few iterations to become familiar before attempting to get zero data on the other loads.

Battlesight Zero. The term battlesight zero means the combination of sight settings and trajectory that greatly reduces or eliminates the need for precise range estimation, holdover or hold-under for your most likely engagements. The battlesight zero is the default setting for you, your gun, your chosen ammunition, and sight/optic combination. A good battlesight zero allows you to accurately engage targets out to a set distance (according to the zero distance you select) without an adjusted aiming point. Common battlesight zeros include 25m, 36yd, 100yd, the 50/200 (can be deceptive) zeros. The key to choosing a zero distance is understanding your requirements (typical engagement ranges in your area) and understanding the trajectory for a given zero. Do not marry yourself to a particular community zero, using a ballistic program and your weapon/ammo combination you may find you need a "non-standard" zero. Standard zeros are great for large organizations with identical weapon systems (Army, USMC). Apply the math for your weapon/ ammo selection to your tactical requirements and dial your system in for *your* requirements. Some zeros will allow you to have three or four inches of difference between line of sight and bullet path out to 200 meters while a different zeroing distance will cause the trajectory /line of sight difference (max ordinate) to be 12-18 inches for the same distances. Each zero answers a different requirement, just make sure it fits yours.

A specific process should be followed when zeroing. The process is designed to be time-efficient and will produce the most accurate zero possible. The zero process includes mechanical zero, grouping and zeroing, and actual zero confirmation out to your zero distance (where the trajectory crosses line of sight the second time). *Do not skip this confirmation step after you zero!* Confirming the actual zero will build confidence in your system / zero and it will also reveal any potential issues.

Fire your first group (can be 3 or 5 rounds, your call but we recommend 5). You can use a laser boresight before the first group to increase the odds of being on paper, just ensure you remove the device before firing. If you are on paper with the first group begin adjusting your optic or iron sights to move the groups to the center of the target (if the group is not on paper move the target closer and repeat the process until you are on then move it back to your zero target distance and repeat). *Avoid the temptation of applying "Kentucky windage" and adjusting your point of aim to account for adjustments.* Keep your point of aim dead center of your zero target during the entire zeroing process. Check your optic or gun's tech manual to determine how many clicks you need at what distance to move the next group...and adjust them in the correct direction. For most modern optics it is simple, moving the knob or slot in the direction of the letter "R" or "U" will move the group in that "right" or "up" direction. And yes, you will eventually make an adjustment in the wrong direction. Everyone makes this mistake at least once, if you see the last group moved the wrong way just double the first correction in the opposite direction, laugh at yourself, and shoot another group. During the zeroing process attempt to center your groups as much as possible, mark and number each group on your target before you fire the next one.

Record the zero data. Once you have fired and adjusted however many groups you need to zero (all rounds from the group inside the inner zero target) add the total number of adjustments from the original mechanical zero. Check your manual for optic and or sights to find out the method to set the indicator markings (if applicable) to "0" for vertical (elevation) and horizontal (windage) turrets. Record the zero data along with round count for that firearm. Replace the target with a fresh one and fire 10 (ten) rounds at the zero target or a B8. If you are satisfied with the zero move on to confirmation.

Confirm the zero. Ranges can be limited for doing so, ideally the zero will be confirmed at the actual zero distance. For example if you use a 200 meter (or 200 yard if you choose that) zero you should confirm at 200 on paper. Second best practice would be to confirm at 200 on steel if paper is not an option at distance (again range restrictions may dictate).

Trust your zero. Once you have a good zero...trust it. Trust yourself. If you start to miss or start tracking to one side of paper targets in a class or during self-directed practice do not try and Kentucky windage it. It is a cycle that will cause you to start chasing rounds and your groups then overall shooting will fall apart. You have either had a mechanical issue or conditions have changed enough to affect your original zero. At the earliest opportunity shoot a fresh group on paper at the first crossover distance (aka the near zero) for your original zero to confirm or deny your suspicions. Re-zero if necessary.

Trust but verify. Whenever you do classes or training in a new location, you should confirm the zero on your rifle since the elevation (altitude), barometric pressure, and other factors may affect the trajectory of a round. There are multitudes of factors that can affect a zero, and the only sure way to know where the rounds are going is to fire the rifle to confirm. The zero on one of your rifles will not transfer to another rifle. This is common sense, but some may think the zero applies to you and will work for all similar guns. It applies to that specific rifle, with that specific ammo only.

Know your holdovers. For every type round (velocity, bullet manufacturer and weight etc) zeroed you need to know your holdovers. A holdover is the height (elevation) you need to hold over a target that is at a distance past your battlesight zero. These will be different for bullet weights and specific loads (cartridges). There are also hold *unders* as well; these will occur between the first point where the bullet crosses the line of sight or "near zero" and the "far zero" distance where the bullet path crosses your line of sight again as it continues to drop. A hold under is when you must hold below the intended target because the path of the bullet is higher than the line of sight at a given point in the trajectory. This data is available in tables and ballistics programs, however it is recommended that you do the work to confirm by shooting known distances on paper with your own rifle. You should have the holdover data for the three types of rounds we discussed at a minimum (your typical training ammo, XM193, and your go to war ammo). It would be prudent to also have holdovers recorded for the ammo that is typical for your team's 5.56 SPR/DMR shooters. If you like using and understand yards and range estimation, then use yards. If you are used to meters, then use meters for zero and holdover data. Just understand why you are using either system and commit to one.

Secondary (Pistol) Secondary weapons are typically our primary in polite society. This can and will change as conditions deteriorate and the pistol fades in importance for most tasks. Carrying or adding a secondary to your kit is driven by METT-TC and your group's SOPs. You may not necessarily carry a secondary for all missions or even at all. This is a thought that may cause issues for some, do some tradeoffs and see if it makes sense. Just don't be afraid to make informed choices. If you feel you need a secondary then plan for carrying one with your setups or plan for addition based on the mission. We won't dive into the secondary weapon or TTPs in this manual, we will keep our focus on the rifle/carbine at this stage.

Firearm Employment in a team environment. A weapons control status (WCS) is a tactical method of fire control given by a leader that incorporates the tactical situation, rules of engagement, and expected or anticipated enemy contact. METT-TC will determine what WCS is at any given time. Why does a Citizen need to know this? Tactical leaders use WCS to control fires aligned with mission parameters to prevent fratricide, abide by current rules of engagement (or laws), and mitigate being compromised during missions. The WCS outlines the target identification conditions under which friendly elements may engage a perceived threat with direct fire and are coupled to any Rules of Engagement (if established).

WEAPONS CONTROL STATUS	DESCRIPTION
WEAPONS HOLD	Engage only if engaged or ordered to engage.
WEAPONS TIGHT	Engage only if target *is positively identified* as *enemy.*
WEAPONS FREE	Engage targets *not positively identified* as *friendly.*

The table above provides a description of the standard WCS used during tactical operations, both in training and combat. They describe when the shooter is authorized to engage a threat target once the threat conditions have been met. You may be given these directions, or as a team or small unit leader you will be responsible for implementing WCS to control your team's weapon status. Understanding these three Weapons Control Status and when they are in place during an operation is critical all the way down to the individual level.

Weapons Hold. Used when absolute fire control over your team is necessary; restrictive ROE (Rules of Engagement), political and civil considerations, or an unclear enemy situation would all call for this status. When conducting surveillance of an objective you will most likely be placed on Weapons Hold (engage only if engaged). It is the most restrictive status that a leader can use.

Weapons Tight. This is the most often used status, think of it as a CCW carrier perspective. You will not engage unless a target is positively identified as a threat.

Weapons Free. Least restrictive, used in high threat, low fratricide opportunity conditions.

Target Prioritization. When faced with multiple targets you must rapidly prioritize and accurately engage each threat. Mental preparedness and the ability to make split-second decisions are the keys to a successful engagement of multiple threats. An offensive combat mindset and prior training (both flat range and scenario based force on force) will allow you to react instinctively and control the pace of the fight, rather than reacting to the threat. All of these prioritization assessments below are near instantaneous, this is not a drawn out deliberate process.

Most dangerous. A threat that has the capability to defeat the friendly force and is preparing to do so. These targets must be defeated immediately.

Dangerous. A threat that has the capability to defeat the friendly force, but is not prepared to do so. These targets are defeated after all most dangerous targets are eliminated. This may also include enemy lethality (machine gun crews maneuvering into position etc), leadership, and communications. You may decide these are Most Dangerous threats (situation dependent), just because they aren't pointing a weapon at you doesn't necessarily mean they aren't the most dangerous.

Least dangerous. Any threat that does not have the ability to defeat the friendly force, but can coordinate with other threats that are more prepared. These targets are defeated after all threats of a higher threat level.

When multiple targets of the same threat level are encountered, the targets are prioritized according to the threat they represent. The standard prioritization of targets establishes the order of engagement. We will engage similar/like threats by:

Near before far.
Frontal before flank.
Stationary before moving.

The prioritization of targets provides a control mechanism for the shooter and facilitates maintaining overmatch against the presented threats. The baseline target prioritization is simplistic and while a good start point we should be prepared to deviate from the prioritization guide. Based on the situation and direction from our leadership these engagement priorities may be adjusted during the Operations Order (OPORD) (eg engage UAS system operators first, engage threat leadership etc) or priorities can be controlled through team fire commands.

Rates of Fire

You must determine how to engage the threat, on the current shot as well as all subsequent rounds. Following the direction of the team leader and your role in the fire team or squad, you will control your rate of fire based on your assessment and your small unit leader's direction.

Rapid Semi Automatic Fire from an AR family of weapons is approximately 45 -60 rounds per minute. Rapid rate of fire places a higher volume of fire on an enemy position is typically used for multiple targets or combat scenarios where the you do not have overmatch of the threat. It is well below the AR (M4) cyclic rate (700-900 rounds per minute and irrelevant for our purposes). Rapid fire consumes ammunition faster than sustained fire and is for short duration only due to potential weapon overheating. The typical GPR or SPR rifleman's firearm (AR platform) will not be capable of these higher rates or volumes of fire for a long period of time.

Slow Semi Automatic Fire (Sustained Fire). Once the enemy has been suppressed, sustained fire conserves ammunition, however it might not be enough volume to suppress or destroy an enemy. Slow semiautomatic fire is moderately paced at the discretion of the individual shooter. This type of firing provides the most time to focus on the functional elements in the shot process. Sustained rates of fire for quality AR pattern carbines and rifles is generally 12 to 15 rounds per minute (check your specific technical manual for your firearm). This does not seem like a lot of rounds per minute, however this is the actual rate that you can sustain over time without overheating to the point of damage or totally destroying your firearm.

Suppressive fire is fire that degrades the performance of a threat below the level needed to fulfil its mission. Direct fire suppression is usually only effective for the duration of the fire. Gaining fire superiority in a fight quickly (and first) is critical, you must be proficient in adjusting your rate of fire to make the best use of every round for the situation and keep every gun in the fight. Knowing when to pick up the firing tempo or slow it down is a skill that must be developed at the team level. Leaders must manage rates of fire, train their teams, and have fire control measure SOPs in place to lead their small units.

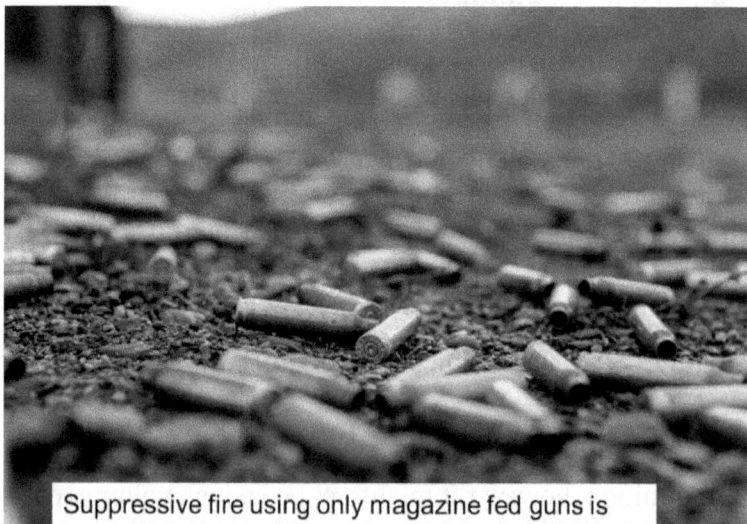

Suppressive fire using only magazine fed guns is tough. Practicing with your teammates is critical. It is not a "mad minute" of mag dumps. Suppression and gaining fire superiority without belt feds is a game of rapid accuracy and synchronization within the team.

Chapter 3
Move

Land Navigation

The Land Nav section of this manual provides the basic knowledge to get your started or refresh your nav skills. Land navigation is a complex skillset that requires a blend of science and art and consists of two methods, Dead Reckoning and Terrain Association. For additional development after these fundamental skills are sharpened a detailed study of the Citizen Land Navigation manual will be required. At entry level you should be proficient in basic map reading (locate a point on a map by 6 or 8 digit grid), identify terrain features on a map, and dead reckoning from one point to another dismounted. Terrain association (the second method of navigation), terrain analysis, detailed route planning, intersection, resection, and numerous other land nav skills are not discussed in detail in this manual. Land navigation is a skill set which takes time and practice to become proficient and confident. Basic mastery of the tools of land navigation -- map, compass, and protractor -- is necessary for mission success. Land navigation requires the use of many tools; the more experienced you become, the more tools you will use. Examples of the basic tools available to the Citizen are the protractor, map, compass, and pace count and arguably small GPS devices (much more at risk to disruption than the other tools). Less obvious examples, that will take experience to appreciate, are the terrain, sun, stars, the direction water flows, wildlife, etc.

The Basic Tools

- Maps
- Compass
- Pencil (.7mm mechanical works best)
- Map Markers
- Waterproofing/Map Case
- Pace Beads
- Notebook
- Protractor
- Straight Edge (can be the protractor)

The Map. A map is simply a graphic representation of a portion of the earth's surface drawn to scale, as seen from above. It uses colors, symbols, and labels to represent features found on the ground. There are numerous types of maps, we will focus our attention on the topographic (or "topo") map for our purposes. It is also wise to have current (within a couple of years) road atlases, state road maps, and specific city maps for your area.

The Topographic Map. A topographic map provides information on the existence, the location of, and the distance between ground features, such as populated places and routes of travel and communication. It also indicates variations in terrain, heights of natural features, and the extent of vegetation cover (to a degree). This type is a map that portrays terrain features in a measurable way (usually through use of contour lines), as well as the horizontal positions of the features represented. The vertical positions, or relief, are represented by contour lines on topographic maps. On maps showing relief, the elevations and contours are measured from a specific vertical datum plane, usually mean sea level.

Having current sets of paper topo maps on hand is critical. The ability to download or acquire map sheets after a potential X Hour could be severely limited. Electronic access to maps during a full-blown crisis will require connecting to networks that may be down while using devices that are sending electronic signatures that could compromise your position or mission.

Current USGS survey maps are available for download for free at the link below:

https://store.usgs.gov/#TM

You can also choose to purchase full sheet maps directly from USGS. They are high quality and are a good value at the time of this manual's printing.

The USGS topobuilder link below is also very useful. It will allow you to customize a 7.5 minute map (1:24,000) and download any area you choose in the US:

https://topobuilder.nationalmap.gov/

Customizing maps of anticipated AOs is a great way to augment the standard USGS quadrangle maps. The website will allow you to choose the area and adjust the layers to your liking before finalizing the custom map that will be emailed to you in a hi res pdf. Removing some layers as you customize it could make the map more useful to you and your team, play around with it and see what works best. With full web access now is the time to allocate a couple of hours to gathering and customizing what you need and it is a great way to become familiar with your local area terrain. There may be things near you that you were unaware of such as abandoned mines, sinkholes, old rail spurs etc. Things that can become incredibly important to you in the future.

All maps after use should be considered sensitive items. If a map falls into unauthorized hands, it could easily endanger operations by providing information of friendly plans or areas of interest to the enemy. Even more important would be a map on which the movements or positions of friendly teams or units were marked. Having a solid destruction plan in place is critical for OPSEC, maps are documents that must not fall into unauthorized hands.

Care must be taken when using a map since it may have to last a long time. If it becomes necessary to mark directly on a map, the use of a pencil is recommended or laminating and using alcohol markers is the best method.

If the map margins must be trimmed for any reason, it is essential to note any marginal information that may be needed later, such as grid data and magnetic declination.

Setting Up a New Map. Preparing a new map for use is simple and will give you an opportunity to get familiar with both the type and data on the particular map sheet. If you are laminating your map, take all these steps prior to laminating as these will be permanent improvements that

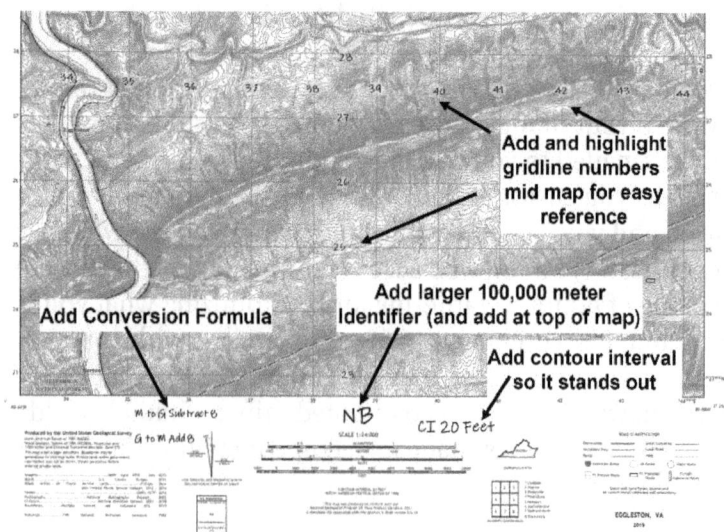

Add and highlight gridline numbers mid map for easy reference

Add larger 100,000 meter Identifier (and add at top of map)

Add Conversion Formula

Add contour interval so it stands out

M to G Subtract B
G to M Add B

NB

CI 20 Feet

EGGLESTON, VA

you will need anytime you use the sheet. These additions are all about convenience and ease of use. Any small modification that will allow you to find data faster, prevent errors, and assist your personal nav preferences should be added.

Grids. Knowing how to determine and report positions on the ground to another team, unit or command post is critical. The grid system is the shared language that gives us a common reference system we use to pass on accurate information to others.

Knowing where you are (position fixing) and being able to communicate that knowledge is crucial to successful land navigation as well as to the effective employment of combat multipliers/ enablers, directing adjacent units toward or away from an enemy position, and medical evacuation. It is essential for valid target location and handover, accurate reporting of obstacles and various danger areas, and obtaining emergency resupply. Few factors contribute as much to the survivability of your team and to successful mission accomplishment as always knowing where you are. You cannot go reinforce / assist another team or call for assistance if you do not know where you are on the ground.

Reference System. In a city the streets are named and the buildings have numbers, to generally locate the only thing needed is the address. Finding locations in undeveloped areas or in unfamiliar parts of the world and then reporting where a point or activity is can be problematic. To solve this problem a uniform and precise system of referencing was developed.

Geographic Coordinates. One of the oldest systematic methods of location is based upon the geographic coordinate system. By drawing a set of east-west rings (latitude) around the globe (parallel to the equator), and a set of north-south rings crossing the equator at right angles (longitude) and converging at the poles, a network of reference lines is formed from which any point on the earth's surface can be located. Lines of longitude (meridians) run north-south and east-west distances are measured between them. Lines of Latitude run east-west and measure north-south distances. All grid reference systems are based on this approach. To locate a precise point on a topographic map using a grid system we will use a coordinate scale - commonly called a protractor. These coordinate scales can be the military Graphic Training Aid (GTA) type or a commercial version that may have the same or different scales. **Ensure you have the correct scale protractor that matches the map you are using.** Scales will differ on maps and the associated coordinate scales; however the one constant will be the measurements of angles which will always be in degrees (most common) or mils (the outer rings with 0-360 degrees or 0-6400 mils). We will always use degrees for navigation. Military (MGRS system) maps are usually 1:50,000 and 1:25,0000 scale maps with some at 1:100,000. We will focus on 1:24,000 USGS maps since those are most likely what we as Citizens will have acquired during our training and preparation. 1:50,000 ("one to fifty") or 1:25,000 may be available during crisis, but the techniques for the USGS map use will apply to these as well. You will need to acquire a 1:24,000 protractor, the military GTA coordinate scale will not work for USGS maps.

Grid Squares. The north-south and east-west grid lines intersect at 90°, forming grid squares. The size of one of these grid squares on USGS and MGRS maps is 1,000 meters x 1,000 meters (1 square kilometer or 1k).

Locate a Point Using Grid Coordinates
Based on the military principle for reading maps (RIGHT and UP), locations on the USGS or MGRS maps can be determined by grid coordinates. We always read map coordinates RIGHT then UP.

The number of digits in a complete grid represents the
level of precision to which a point has been located and
measured on a map; the more digits the more precise the
measurement. A four digit grid is accurate to 1,000 meters,

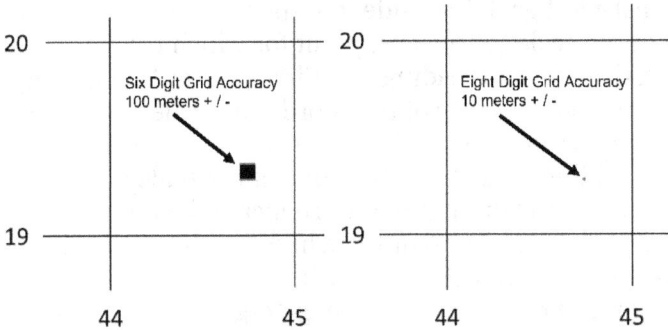

a six digit grid will get us within 100 meters of a location,
an 8 digit grid within 10 meters, and a 10 digit within 1
meter (typically 10 digit grids are GPS, paper map and
protractor will only get you to 8 digits resolution
realistically). The more errors we can remove (math errors,
pencil or marker thickness, protractor placement on the
map etc) the more precision we will have.

With a Coordinate Scale (1:24,000 scale)
In order to use the coordinate scale for determining grid
coordinates, ensure that the appropriate scale is being used
on the corresponding map and that the scale is right side
up. On the 1:24,000 coordinate scale, there are two axis on
the triangle or backwards "L" shaped scale: the horizontal
(Easting or first reading – read RIGHT) and vertical
(Northing or second reading – read UP). These sides are
1,000 meters in length. The point at which the north/south
and east/west sides of the scale meet (the bottom right
corner of the scale triangle) is the zero-zero point. Each
1,000 meter side is divided into 10 equal 100-meter
segments by a long tick mark and number. Each 100-meter
segment is subdivided into 10-meter segments by a short
tick mark with a mid point or 50 meter line between each

numbered line. To ensure the scale is correctly aligned, place it with the zero-zero point at the lower left corner of the grid square.

Keeping the horizontal line of the scale directly on top of the east-west grid line, slide it to the right until the vertical line of the scale touches the point for which the coordinates are desired. When reading coordinates, stay "squared up" - examine the two sides of the coordinate scale to ensure that the horizontal line of the scale is aligned with the east-west grid line, and the vertical line of the scale is parallel with the north-south grid line. To locate the point to the nearest 10 meters, measure the hundredths of a grid square RIGHT and UP from the grid lines to the point. The point in the figure is in the 4520 grid square (remember grids expressed as numbers are ALWAYS an even number). The coordinates to the nearest ten meters are 45352016. If a six-digit grid was given the grid would be 454202 and would be accurate to 100 meters.

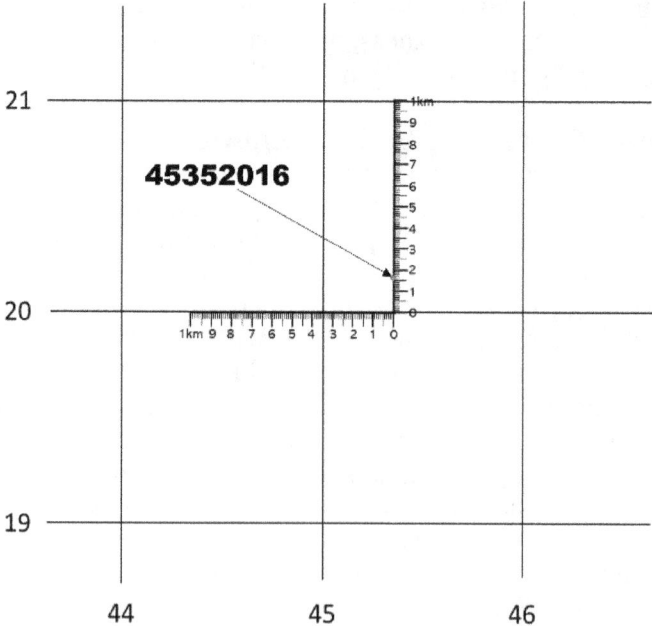

Recording and Reporting Grid Coordinates

Coordinates are written as one continuous number without spaces, parentheses, dashes, or decimal points; grid coordinates *always contain an even number of digits.* Therefore, whoever is to use the written coordinates must know where to make the split between the RIGHT and UP readings. Since there are thousands of duplicate coordinate numbers across the full series of map sheets it is important that the 100,000-meter square identification letters (described below) be included in any point designation. Normally, grid coordinates are determined to the nearest 100 meters (six digit grid) for expedient reporting locations.

100,000 Meter Square Identifier

Every map sheet is part of a larger grid system. Numbers and letters are used to identify the section the map is associated with. This information is found in the marginal information in the map sheet. Further explanation of marginal info in in the CM-5 Citizen Land Navigation manual (again the scope of the land nav section in this manual is intentionally kept minimal). Sometimes organizations will work across map sheet boundaries and to prevent confusion we use the 100,000 Meter Square Identifier at the beginning of the grid coordinate. In this example we would use "NB" or "November Bravo" when referencing grids on this map sheet. An example of a location reported on the radio:

"WARRIOR XRAY this is CALUMET SIX EIGHT, one stationary BMP in the open at NB454202"

The 100,000 meter identifier is mistakenly (even by professionals) referred to as the 'Grid Zone Designator". The Grid Zone Designator is a set of numbers and a letter eg "17S" to further identify where a series of map sheets falls into the larger grid scheme. It is one level above the 100k meter identifier and is seldom used for our purposes other than locating and acquiring the proper map sheets for an area.

A quick note on 1:50,000 Coordinate Scale
The 1:50,000 scale is subtended in 50 meter segments, by using interpolation, mentally divide each 50-meter segment into tenths. For example, a point that lies after a whole number but before a short tick mark is identified as 10, 20, 30, or 40 meters and any point that lies after the short tick mark but before the whole number is identified as 60, 70, 80, or 90 meters. Used in the same manner as the 1:24,000, but the 1:50,000 scale is typically used on MGRS or military maps.

Map Colors. Colors on a standard topographic map are:

1. **Black.** Indicates cultural (man-made) features such as buildings and roads, surveyed spot elevations, and all labels.
2. **Red-Brown**. All relief features, non-surveyed spot elevations, and elevation, such as contour lines on redlight readable maps.
3. **Blue.** Identifies hydrography or water features such as lakes, swamps, rivers, and drainage.
4. **Green.** Identifies vegetation with military significance, such as woods, orchards, and vineyards (these can easily change over time and may be outdated after the map is printed).
5. **Brown.** Identifies all relief features and elevation, such as contours on older edition maps, and cultivated land on red-light readable maps.
6. **Red.** Classifies cultural features, such as populated areas, main roads, and boundaries, on older maps.
7. **Other.** Occasionally other colors may be used to show special information. These are indicated in the marginal information as a rule.

Terrain Features

The topographic map is an accurate representation of the ground, we just have to know how to "read" the map. Learning to read a map is akin to learning a new language, with practice it can become second nature and eventually the map features will make sense as your brain pulls it all together. The lines and shapes that are arranged to represent terrain features will become meaningful. When you look at an area of the map you are planning to traverse and literally say out loud "man this is really going to suck"...you are well on your way to knowing how to read a map.

All terrain features are derived from a complex landmass known as a mountain or ridgeline (see figure below). The term ridgeline is not interchangeable with the term ridge. A ridgeline is a line of high ground, usually with changes in elevation along its top and low ground on all sides from which a total of 10 natural or man-made terrain features are classified.

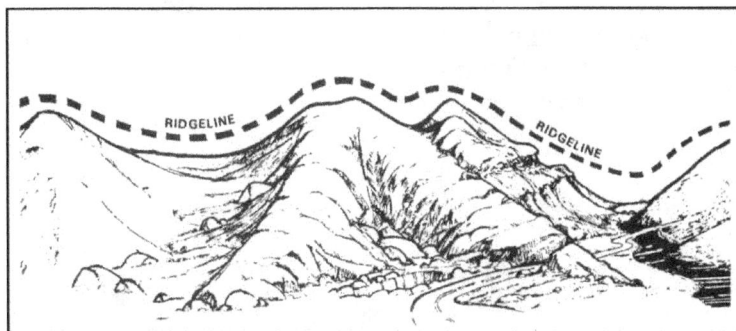

Major Terrain Features (there are five)

Hill. A hill is an area of high ground. From a hilltop, the ground slopes down in all directions. A hill is shown on a map by contour lines forming concentric circles. The inside of the smallest closed circle is the hilltop.

HILL

Saddle. A saddle is a dip or low point between two areas of higher ground. A saddle is not necessarily the lower ground between two hilltops; it may be simply a dip or break along a level ridge crest. If you are in a saddle, there is high ground in two opposite directions and lower ground in the other two directions. A saddle is normally represented as an hourglass.

SADDLE

Valley. A valley is a stretched-out groove in the land, usually formed by streams or rivers. A valley begins with high ground on three sides, and usually has a course of running water through it (solid blue line for persistent water, dashed/dotted line for an intermittent (seasonal or wet weather) stream. If standing in a valley, three directions offer high ground, while the fourth direction offers low ground. Depending on its size and where a person is standing, it may not be obvious that there is high ground in the third direction, but water flows from higher to lower ground. Contour lines forming a valley are either U-shaped or V-shaped. To determine the direction water is flowing, look at the contour lines. The closed end of the contour line (U or V) always points upstream or toward high ground.

VALLEY

Ridge. A ridge is a sloping line of high ground. Standing on the centerline of a ridge, you will normally have low ground in three directions and high ground in one direction with varying degrees of slope. If you cross a ridge at right angles, you will climb steeply to the crest and then descend steeply to the base. When you move along the path of the ridge, depending on the geographic location, there may be either an almost unnoticeable slope or an obvious incline. Contour lines forming a ridge tend to be U-shaped or V-shaped. The closed end of the contour line points away from high ground.

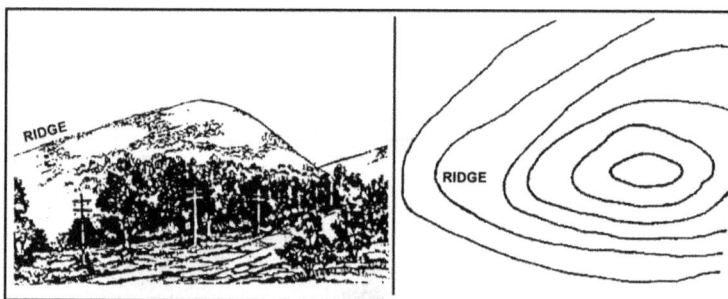

RIDGE

Depression. A depression is a low point in the ground or a sinkhole. It could be described as an area of low ground surrounded by higher ground in all directions, or simply a hole in the ground. Usually only depressions that are equal to or greater than the contour interval will be shown. On maps, depressions are represented by closed contour lines that have tick marks pointing toward low ground.

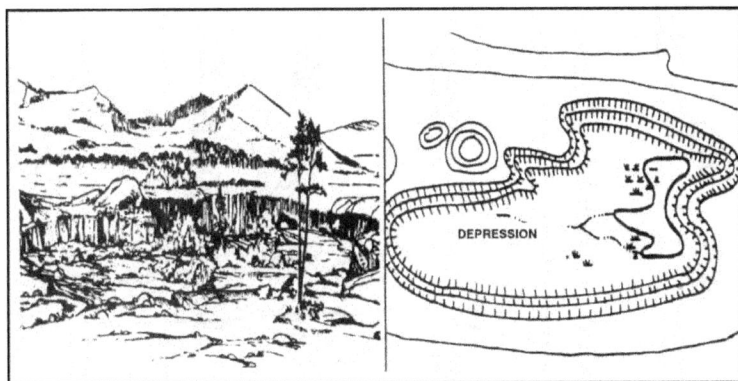

DEPRESSION

Minor Terrain Features (there are three)

Draw. A draw is a less developed stream course than a valley. In a draw, there is essentially no level ground and, therefore, little or no maneuver room within its confines. If you are standing in a draw, the ground slopes upward in three directions and downward in the other direction. A draw could be considered as the initial formation of a valley. The contour lines depicting a draw are U-shaped or V-shaped, pointing toward high ground.

DRAW

Spur. A spur is a short, continuous sloping line of higher ground, normally jutting out from the side of a ridge. A spur is often formed over time by two rough parallel streams (water may or may not be present today) , which cut draws down the side of a ridge. The ground sloped down in three directions and up in one direction. Contour lines on a map depict a spur with the U or V pointing away from high ground.

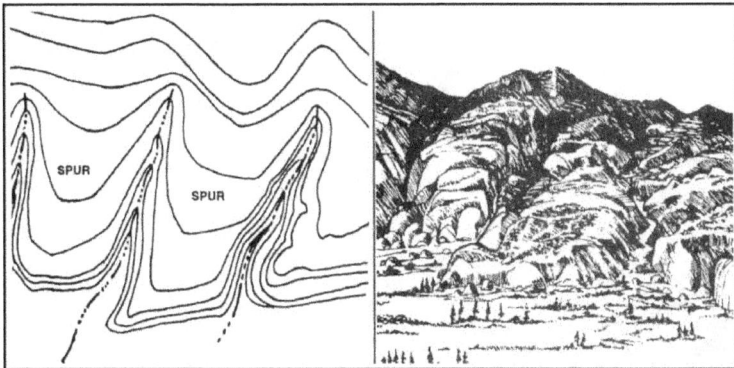

SPUR

Cliff. A cliff is a vertical or near vertical feature; it is an abrupt change of the land. When a slope is so steep that the contour lines converge into one "carrying" contour of contours, this last contour line has tick marks pointing toward low ground. Cliffs are also shown by contour lines very close together and, in some instances, touching each other. A critical terrain feature that everyone in your team needs to be aware of when maneuvering near it.

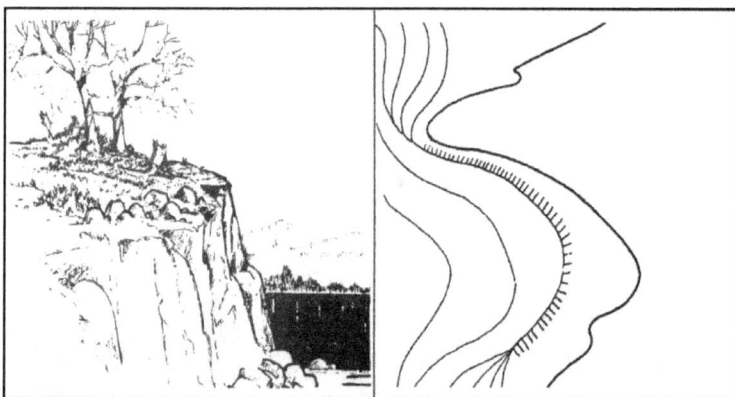

CLIFF (Defined)

CONVERGING
CONTOURS
FORMING CLIFF

CLIFF (Indicated by Contour)

Supplementary Terrain Features (there are two)

Cut. A cut is a man-made feature resulting from cutting through raised ground, usually to form a level bed for a road or railroad track. Cuts are shown on a map when they are at least 10 feet high, and they are drawn with a contour line along the cut line. This contour line extends the length of the cut and has tick marks that extend from the cut line to the roadbed, if the map scale permits this level of detail.

Fill. A fill is a man-made feature resulting from filling a low area, usually to form a level bed for a road or railroad track. Fills are shown on a map when they are at least 10 feet high, and they are drawn with a contour line along the fill line. This contour line extends the length of the filled area and has tick marks that point toward lower ground. If the map scale permits, the length of the fill tick marks are drawn to scale and extend from the base line of the fill symbol.

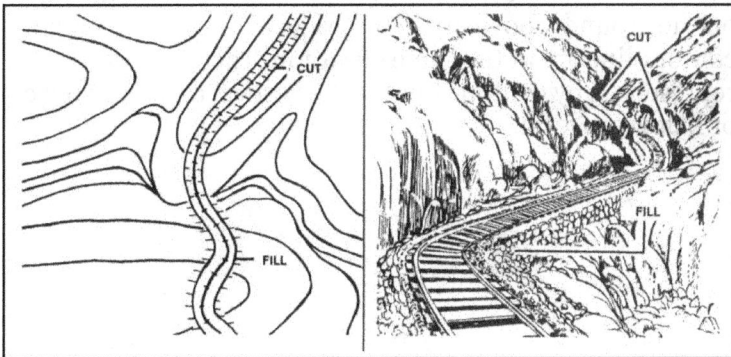

CUT and FILL

Elevation and Relief

The elevation of points on the ground and the relief of an area affect the movement, positioning, and, in some cases, effectiveness of your team. You must know how to determine locations of points on a map, measure distances and azimuths, and identify symbols on a map. Being able to determine the elevation and relief of areas on standard topo maps is a required skill for Pro Citizens. To do this, you must first understand how the mapmaker indicated the elevation and relief on the map.

Elevation of a point on the earth's surface is the vertical distance it is above or below mean sea level.

Relief is the representation (as depicted by the mapmaker) of the shapes of hills, valleys, streams, or terrain features on the earth's surface.

Methods of Depicting Relief

Mapmakers use several methods to depict relief of the terrain. The most common and the one we will focus on are *Contour Lines.* A contour line represents an imaginary line on the ground, above or below sea level. All points on the contour line are at the same elevation. The elevation represented by contour lines is the vertical distance above or below sea level.

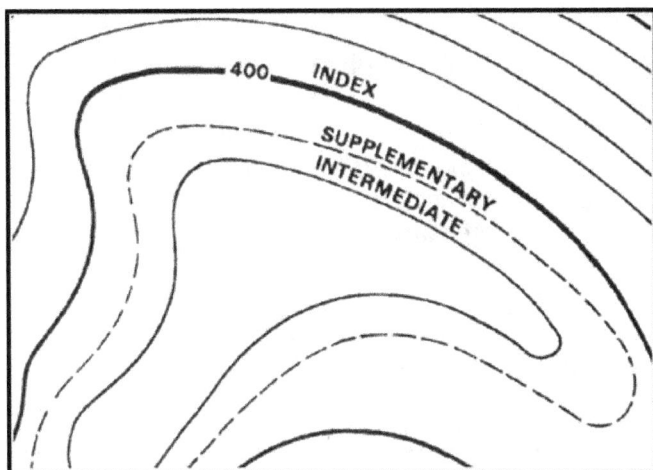

The three types of contour lines used on a standard topographic map are:

Index. Starting at zero elevation or mean sea level, every fifth contour line is a heavier line. These are known as index contour lines. Normally, each index contour line is numbered at some point. This number is the elevation of that line.

Intermediate. The contour lines falling between the index contour lines are called intermediate contour lines. These lines are finer and do not have their elevations given. There are normally four intermediate contour lines between index contour lines.

Supplementary. These contour lines resemble dashes. They show changes in elevation of at least one-half the contour interval. These lines are normally found where there is very little change in elevation, such as on fairly level terrain.

Contour Intervals

Before the elevation of any point on the map can be determined, the user must know the contour interval for the map he is using. The contour interval may be in meters or in feet. The interval measurement given in the marginal information is the vertical distance between adjacent contour lines. To determine the elevation of a point on the map:

Determine the contour interval and the unit of measure used, for example, feet, meters, or yards and find the numbered index contour line nearest the point of which you are trying to determine the elevation.

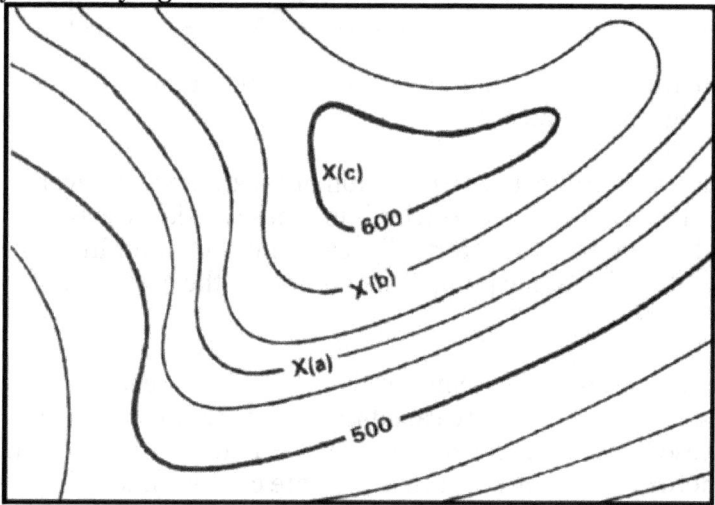

Determine if you are going from lower elevation to higher, or vice versa. In the figure above point (a) is between the index contour lines. The lower index contour line is numbered 500, which means any point on that line is at an elevation of 500 meters above mean sea level. The upper index contour line is numbered 600, or 600 meters. Going from the lower to the upper index contour line shows an increase in elevation.

89

To determine the exact elevation of point (a), start at the index contour line numbered 500 and count the number of intermediate contour lines to point (a). Locate point (a) on the second intermediate contour line above the 500-meter index contour line. The contour interval is 20 meters, thus each one of the intermediate contour lines crossed to get to point (a) adds 20 meters to the 500-meter index contour line. The elevation of point (a) is 540 meters; the elevation has increased.

To determine the elevation of point (b) find the nearest index contour line. In this case, it is the upper index contour line numbered 600. Locate point (b) on the intermediate contour line immediately below the 600-meter index contour line (below means downhill or a lower elevation). Point (b) is located at an elevation of 580 meters. If you are increasing elevation, add the contour interval to the nearest index contour line. If you are decreasing elevation, subtract the contour interval from the nearest index contour line.

To determine the elevation to a hilltop at point (c) add one-half the contour interval to the elevation of the last contour line. In this example, the last contour line before the hilltop is an index contour line numbered 600. Add one-half the contour interval, 10 meters, to the index contour line. The elevation of the hilltop would be 610 meters.

In addition to the contour lines, bench marks and spot elevations are used to indicate points of known elevations on the map.

Bench marks are the more accurate of the two, are symbolized by a black X, such as X BM 214. The 214 indicates that the center of the X is at an elevation of 214 units of measure (feet, meters, or yards) above mean sea level.

Spot elevations are shown by a brown X and are usually located at road junctions and on hilltops and other prominent terrain features. If the elevation is shown in black numerals, it has been checked for accuracy; if it is in brown, it has not been checked.

NOTE: New maps are being printed using a dot instead of brown Xs

Interpretation of Terrain Features

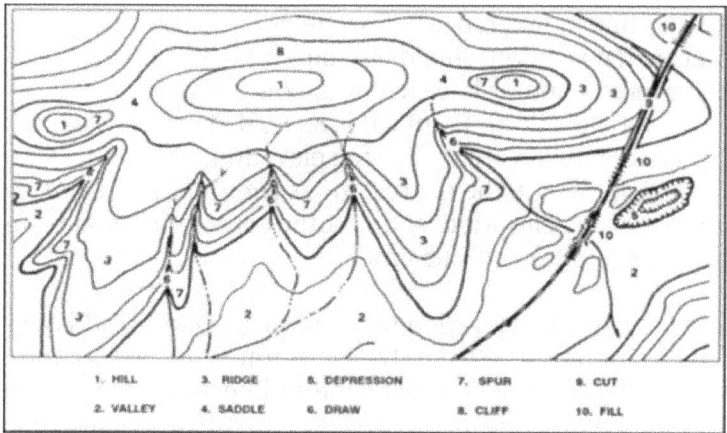

| 1. HILL | 3. RIDGE | 5. DEPRESSION | 7. SPUR | 9. CUT |
| 2. VALLEY | 4. SADDLE | 6. DRAW | 8. CLIFF | 10. FILL |

TERRAIN FEATURES IN RELATION

Terrain features do not stand alone, it is the relationship on the ground that we need to understand to be a proficient navigator and be combat effective. To better understand these as depicted on a map and effectively plan routes and visualize maneuver corridors, you need to interpret them. These methods are discussed in the CM-5 Citizen Land Navigation Manual and the other Small Unit Tactics references in the Professional Citizen series. Foundational terrain analysis (the terrain features in relation coupled with mission variables) is done by using the acronym OAKOC.

OAKOC

Military aspects of terrain OAKOC are used to analyze the ground and determine the effects of each aspect of terrain on both friendly and enemy forces. We do this to determine who has the advantage and more importantly to figure out how to gain an advantage over the enemy. If they have excellent cover and concealment in a static defensive position we may be able to mitigate this advantage by moving at night, in poor weather conditions over routes that are outside of their direct fire range. The analysis always has a purpose, it is not to recite the acronyms or build out a checklist. The terrain analysis whether deliberate or on the run is done to provide us with information to make decisions that will give us an advantage over the enemy. As you grow and become a leader these terrain effects translate directly into planning assumptions applying to friendly or enemy Course of Action (COA) for small unit leaders. At the individual Citizen level you will most likely use terrain analysis to help decide your movement routes, position an observation post (OP), or to select your own individual firing position for example. We should allocate as much time as possible during mission analysis to analyze terrain and weather aspects, however you should be able to rapidly analyze a piece of terrain you are on as you move or as you take up a temporary support or observation position.

Always look at / analyze terrain from both perspectives (enemy AND friendly) in a three-dimensional space to include UAS observation and electronic line of sight for communications masking

This analysis allows us to template where enemy forces may be, position our own assets for security or defensive purposes, select movement techniques and formations, and informs timing and location of employ combat multiplier (aka "force multiplier") employment (unmanned systems for example).

OAKOC - Obstacles. You will identify existing (inherent to terrain and either natural or man-made) and reinforcing obstacles limiting mobility in your AO. Reinforcing obstacles are constructed, emplaced, or detonated by a military or guerilla force.

Existing obstacles include things that may impede movement (mounted or dismounted). Natural existing obstacles include rivers; forests; mountains; ravines; gaps and ditches, tree stumps and large rocks, forests that can impede off road vehicular movement. Remember you are looking at this from both the friendly and enemy perspective. Just because our vehicles may not be able to overcome the obstacle does not mean an enemy tracked vehicle would be stopped. Existing obstacles of the man-made type include towns, canals, railroad embankments, buildings, power lines etc. These obstacles are things we see in daily life that we can mentally train ourselves to assess. Wargaming and mentally doing a "what if?" drill is an easy and effective way to develop this skillset as you drive through urban areas, neighborhoods, and over bridges. Where would you defend that piece of terrain using the existing obstacle? How would you bypass or negotiate it if you were on the offense?

Reinforcing obstacles are emplaced to inhibit the ability of the opposing force to move, mass, and reinforce. Examples include mine fields, antitank ditches, abatis, and concertina wire obstacles / emplaced fencing.

OAKOC - **Avenues of Approach** is the anticipated or templated routes or groups of routes. An Avenue of Approach* is an air or ground route of an attacking force leading to an objective or key terrain. Avenues of approach are classified by type (mounted, dismounted, air, or subterranean), formation, and speed of the largest unit traveling on it. We must determine if variations in trafficability force changes in our formations or movement techniques, consider advantages and disadvantages of each avenue, and the enemy's likely counterattack routes. Again looking at it from the enemy perspective determine the likely enemy avenues into our AO, how the enemy can use each avenue of approach and what lateral routes could the enemy use to threaten our flanks?

friendly forces are not referred to as having Avenues of Approach; we use Routes or an Axis of Advance when referencing friendly maneuver graphic control measures after our routes have been planned.

OAKOC - **Key Terrain** is location(s) or areas whose seizure, retention, or control gives a marked advantage to either combatant. It is a conclusion, usually arrived at after enemy analysis and COA development, rather than an observation. A prominent hilltop overlooking an avenue of approach might or might not be key terrain. Even if it offers clear observation and fields of fire, it offers nothing if the enemy can easily bypass it, or if the selected course of action involves maneuver on a different avenue of approach. Decisive terrain is one "level up" from Key Terrain; it is always Key Terrain but the reverse is not true. Decisive terrain's seizure, retention, or control is necessary for mission accomplishment, it will most likely determine the outcome of the fight. Some situations have no decisive terrain. This concept is covered in greater detail in our other manuals in this series.

OAKOC - **Observation and Fields of Fire** are identified locations along each avenue of approach providing clear observation and fields of fire for both the attacker and defender. Analyze the area surrounding key terrain, objectives, engagement areas, obstacles and locate intervisibility lines (ridges or horizons which can hide equipment or personnel from observation). You must assess the ability of the attacking force to overwatch or support their (or your if you are the attacker/recon force) movement with direct fire. Identify positions where enemy artillery or mortar observers can call for indirect fire against your team. When possible conduct a ground reconnaissance from both enemy and friendly perspectives. You might do it personally, by map, or with friendly locals providing information, or use recent satellite imagery if available. When on the offense determine where / if there are clear observation and fields of fire available on or near the objective for enemy observers and weapon systems and assess where the enemy will and will not be able to concentrate their fires. Template the location of enemy vulnerabilities (flanks, unit boundaries, dead space etc). Defensive considerations include determining locations with clear observation and fields of fire along enemy avenues, assessing where the enemy will establish support by fire positions, and finding the dead space in your area of operations to determine where you are vulnerable.

OAKOC - **Cover and Concealment** is mistakenly confused by many, they are not interchangeable terms. Cover is natural or man-made ballistic protection from the effects of direct and indirect fire being used against a force. Cover may change and become ineffective at any time as new weapons systems are brought to bear against it. Concealment is simply protection from observation and provides no relevant ballistic protection. Consider the terrain, vegetation, structures, and other features along avenues of approach (and on objectives or key terrain) to identify sites offering good or poor cover and concealment. In the defense, weapon positions must be both lethal to the enemy and survivable to the Citizen. This element must be considered just as all the others, look at it from both perspectives and determine who has the advantage and why.

Categories of terrain

When analyzing terrain for ground maneuverability during the IPB Process (Intelligence Preparation of the Battlefield) the terrain is categorized into one of three categories for analysis. Very briefly this just indicates the difficulty of maneuvering (accounting for being mounted and dismounted) through a given series of terrain. *Unrestricted* terrain is easy, *Restricted* will slow your movement (think thick vegetation, steep terrain, or passable swamp), and *Severely Restricted* is near impossible without some type of engineer, water craft, or air mobility assistance. For Severely think cliffs, flooded areas, etc. Determining and assessing these during the IPB process are not addressed in this manual however you will likely hear these terms as you participate in training and will read about them in the other Citizen Manuals.

Terrain Related Terms You May Hear and Eventually Use

Some of the terms below are doctrinal, some are tacit / institutional knowledge and though unofficial they are part of our common language.

Go, Slow-Go, No-Go (Terrain). In military terms, "go", "no-go" and "slow-go" terrain were used to describe the maneuverability of the terrain. These have been replaced with *"Unrestricted, Restricted, and Severely Restricted"* however you may still hear the older terms (incorrectly) used.

Low Ground. Terrain that is lower / has less elevation than the surrounding terrain. Low ground typically has denser vegetation, swampland, water etc and is usually at a tactical disadvantage compared to surrounding high ground.

Intervisibility Line (IV Line). Intervisibility Lines (IVLs) are not terrain features and do not appear on maps as discrete entities. IVLs are an effect of terrain on line of sight / observation. Terrain features can be readily identified regardless of the observer's location, IVLs are only relative to the observers perspective (elevation and direction). IV lines are a localized pattern of limitations on observation caused by often subtle variations in terrain elevation, aka folds in the ground. Picture a long straight two lane road. Oncoming vehicles may seem to disappear in spots as they approach. These are IV Lines. IV lines can provide a significant advantage to the side who knows the terrain; a proficient unit or team can infiltrate laterally along concealed IV lines.

Danger Area. The term "danger area" refers to any area on the route where the terrain would expose you or your team to enemy observation, fire, or both. Examples include large open areas, roads and trails, and bridges or crossing sites over water obstacles.

Linear Danger Area. A linear danger area is an area where the team or unit's flanks are exposed along a relatively narrow field of fire. Examples include building lined streets, tree lined roads, narrow trails, and streams.

High Speed Avenue (of Approach). An unrestricted mobility corridor that affords a mounted enemy access to or through a piece of terrain. Hardball roads with little restrictive terrain or canalization are typically high speed avenues of approach.

Military Crest. An area on the forward or reverse slope of a hill or ridge just below the topographical crest from which maximum observation and direct fire covering the slope down to the base of the hill or ridge can be obtained.

Hardball / Hardball Road. A paved road.

Goat Trail. A very narrow trail through rough or difficult terrain, usually but not always single track. Some will call any tank trail or dirt road a goat trail. These people would be what we call "wrong".

Defile. A defile is a narrow pass or gorge between mountains or hills. The term originates from a military description of a route through which troops can march only in a narrow column or with a narrow front. A defile may be defined by a narrow valley, draw, or cut (all terrain features) or by vegetation, man made structures or geographic features that may not be captured on your map as terrain features.

Front Line Trace. Forward location of a unit conveyed in relation to a graphic control measure (eg Phase Line RUPE) or along a northing or easting grid line (eg along the 82 Easting or 82 grid line). These are never reported by unsecure means.

The high ground is not always Key Terrain, but it will most always present a tactical dilemma for the force that is occupying low ground. It may become Key Terrain just by virtue of a narrow keyhole line of sight (LOS) into main supply routes and lines of communication.

Using Direction and Azimuths to Navigate

Direction. Directions are expressed as units of angular measure. The method we use to express direction that is accurate, is adaptable to any part of the world, and has a common unit of measure is the Degree. A degree is the most common unit of measure is the degree (°) with its subdivisions of minutes (') and seconds ("). 1 degree = 60 minutes, 1 minute = 60 seconds. You may see or hear Mil or Grad as units of measure as well, but for our purposes we will use Degrees.

Base Lines. In order to measure something, there must always be a starting point or zero measurement. To express direction as a unit of angular measure, there must be a starting point or zero measure and a point of reference. These two points designate the base or reference line.

There are three base lines: true north, magnetic north, and grid north. The most commonly used are magnetic and grid north.

True North. A line from any point on the earth's surface to the north pole. All lines of longitude are true north lines. True north is usually represented by a star.

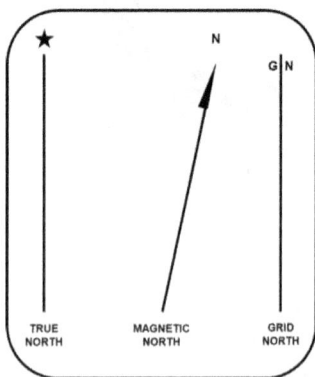

Magnetic North. The direction to the north magnetic pole, as indicated by the northseeking needle of a magnetic instrument. The magnetic north is usually symbolized by a line ending with half of an arrowhead. Magnetic readings are obtained with magnetic instruments, such as lensatic or baseplate compasses.

Grid North. The north that is established by using the vertical grid lines on the map. Grid north may be symbolized by the letters GN or the letter "y".

Azimuth. An azimuth is defined as a horizontal angle measured clockwise from a north base line. This north base line could be true north, magnetic north, or grid north. The azimuth is the most common military method to express direction. When using an azimuth, the point from which the azimuth originates is the center of an imaginary circle. This circle is divided into 360 degrees.

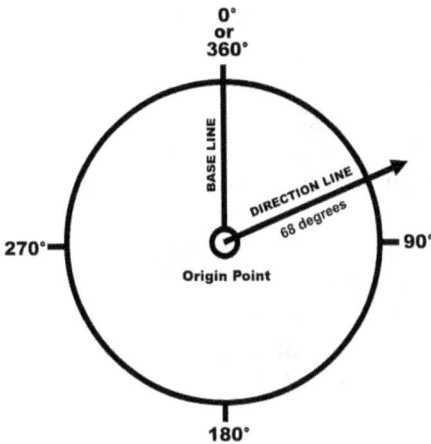

Back Azimuth. A back azimuth is the opposite direction of an azimuth. It is comparable to doing "about face." To obtain a back azimuth from an azimuth, add 180 degrees if the azimuth is 180 degrees or less, or subtract 180 degrees if the azimuth is 180 degrees or more. The back azimuth of 180 degrees may be stated as 0 degrees or 360 degrees.

Grid Azimuth. When an azimuth is plotted on a map between the start point and end point the points are joined together by a straight line (if using a round protractor you will need a dedicated straight edge for accurate map work). Use your protractor to measure the angle between grid north and the drawn line, and this measured azimuth is the **grid azimuth.** When measuring azimuths on a map, remember that you are measuring from the start point to an end point. If you mistakenly take the reading from the end point, the grid azimuth will be opposite causing you or your team to go in the wrong direction.

Protractor. There are several types of protractors—full circle, half circle, square, and rectangular. All of them divide the circle into units of angular measure, and each has a scale around the outer edge and an index mark. The index mark is the center of the protractor circle from which all directions are measured. If you choose a circular protractor be sure to have a straight edge in your kit as well. Often square protractors will have Mils on the outside and degrees on the inner circle. Some navigators will trim off the mils which is more convenient and user friendly. Just be very precise when trimming so you maintain the straight edge. Double or multiple scale coordinate scales are great, just make sure you are using the correct set of numbers when you are cold, wet, and out of your mind tired.

When using the protractor, the base line is always oriented parallel to an Easting (a north-south grid line). The 0- or 360-degree mark is always toward north on the map and the 90° mark is to the east.

To measure / determine a grid azimuth

- Draw a line connecting the two points (start and end).

- Place the index of the protractor at the point where the drawn line crosses an Easting (a north-south grid line) close to the start point grid line.

- Keeping the index at this point, adjust (rotate) the 0- to 180-degree line of the protractor on the Easting.

- Read the value of the angle from the scale; this is the grid azimuth from start point to end point (see diagram below).

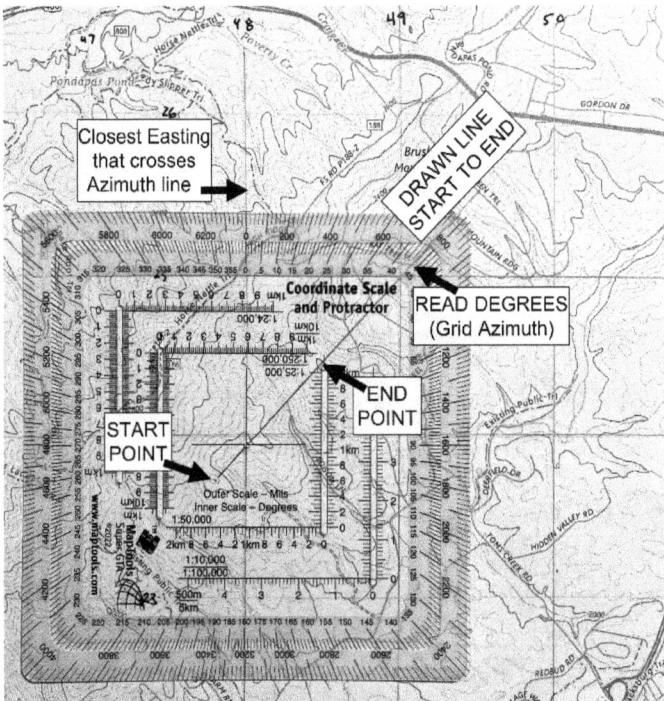

Determine a Grid Azimuth (GAZ)

Plot an azimuth from a known point on a map

- Convert the azimuth from magnetic to grid, if necessary. (See GM angle conversion section later in this manual.)

- Place the protractor on the map with the index mark at the center of mass of the known point and the base line parallel to an Easting.

- Make a mark on the map at the desired azimuth.

- Remove the protractor and draw a line connecting the known point and the mark on the map. This is the grid direction line (azimuth). When measuring an azimuth, the reading is always to the nearest degree, distance does not change an accurately measured azimuth.

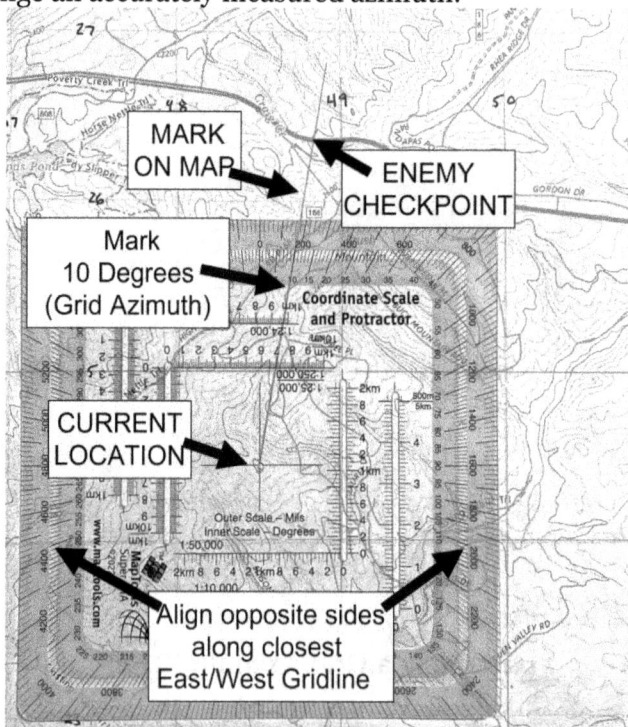

Plot a Grid Azimuth

Square Your Protractor. To obtain an accurate reading the protractor must be squared up to the map. There are two techniques to check that the protractor is aligned to the grids:

(1) Place the protractor index where the azimuth line you drew cuts a north-south grid line, aligning the base line of the protractor directly over the intersection of the azimuth line with the north-south grid line. You can now determine whether the initial azimuth reading was correct.

(2) Note that the protractor is bisected at both the top and bottom by the same north-south grid line (Easting). Count the number of degrees from the 0-degree mark at the top of the protractor to this grid line and then count the number of degrees from the 180-degree mark at the bottom of the protractor to this same Easting. If the two counts are equal, the protractor is properly aligned. The same can be done for an east/west (Northing) gridline and the sides of the protractor.

Compass to Map and Map to Compass. Ok here is where we have to do some math. To account for the difference between the paper map grid system and our compass that points to a magnetic field (that is not at the North Pole) we must add or subtract. The trick is knowing when to add and when to subtract, other than that it is straightforward and easy. The Declination Diagram will give us the known values to add or subtract to our azimuths.

Declination Diagram. Declination is the angular difference between any two norths. If you have a map and a compass, the one of most interest to you will be between magnetic and grid north. Your map is laid out in the projection that uses "Grid North" while your compass will point to the Earth's magnetic north. Magnetic north changes each year and will shift considerably over time. It is important you know the date of the data on your map to ensure you are using relatively new declination diagram data. The declination diagram found in the marginal information of your map shows the angular relationship, represented by prongs, among grid, magnetic, and true norths. While the relative positions of the prongs are correct, they are seldom plotted to scale. **Do not use the diagram drawing to physically measure a numerical value.** This value will be written in the map margin (in both degrees and mils) beside the diagram. True North on the diagram is of no interest to us as land navigators at this point in our progression. For now just ignore it in the Declination Diagram.

Grid-Magnetic Angle. The G-M angle value is the angular size that exists between grid north and magnetic north. It is an arc, indicated by a dashed line, that connects the grid north and magnetic-north prongs. This value is expressed to the nearest 1/2 degree, with mil equivalents shown to the nearest 10 mils (again we will deal with Degrees only for now). The G-M angle is important to us because azimuths translated between map and ground will be in error by the size of the declination angle if not adjusted for it.

Conversion. Since the location of magnetic north does not correspond exactly with the grid-north lines on the maps, a conversion from magnetic to grid or vice versa is needed using the G-M Angle found in the Dec Diagram in the marginal data of our maps. This is where we have to do some simple math to go from Magnetic Azimuth (MAZ)

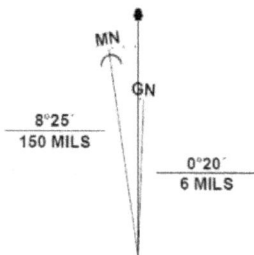

MN

GN

8°25´
150 MILS

0°20´
6 MILS

UTM GRID AND 2021 MAGNETIC NORTH
DECLINATION AT CENTER OF SHEET

to Grid Azimuth (GAZ) or grid to magnetic.

Diagrams with Notes. Simply refer to the conversion notes that appear in conjunction with the diagram explaining the use of the G-M angle. One note provides instructions for converting magnetic azimuth to grid azimuth; the other, for converting grid azimuth to magnetic azimuth. The conversion (add or subtract) is governed by the direction of the magnetic-north prong relative to that of the north-grid prong. USGS will rarely (if ever) have the conversion notes on the Dec Diagram, it is wise to write the formula in when you receive a new map.

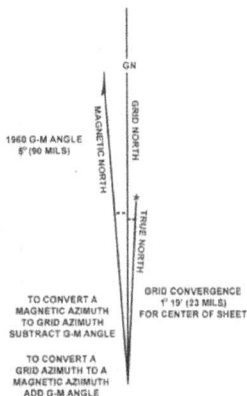

GN

MAGNETIC NORTH

GRID NORTH

TRUE NORTH

1960 G-M ANGLE
6° (90 MILS)

GRID CONVERGENCE
1° 19' (23 MILS)
FOR CENTER OF SHEET

TO CONVERT A
MAGNETIC AZIMUTH
TO GRID AZIMUTH
SUBTRACT G-M ANGLE

TO CONVERT A
GRID AZIMUTH TO A
MAGNETIC AZIMUTH
ADD G-M ANGLE

Converting from One Type Azimuth to Another

The easiest way to wrap your head around this is the diagrams below. Remembering when to add and subtract is based on which side of the declination diagram your map sheet is on. The Western US has East Declination (Mag North Arrow is to the right (East) of the Grid North line in the Declination Diagram) while the Eastern US has a West Declination (Mag North Arrow is to the left (West) of the Grid North line in the Declination Diagram). Remember the amount of declination (either direction) moves gradually over the years as magnetic north drifts and shifts.

Math Note: Remember, there are no negative azimuths on the azimuth circle. If your Azimuth you are converting from is less than the G-M angle (and you are using one of the two subtraction formulas) you will have to add 360 before you do the conversion math. Since 0 degrees is the same as 360 degrees, then 7 degrees is the same as 367 degrees. 7 degrees and 367 degrees are located at the same point on the azimuth circle. The small number azimuth can now be converted after adding 360 because it is now larger than the G-M angle (use only when subtracting).

If the Declination Diagram Looks Like This (Mag North is to the East of Grid North)

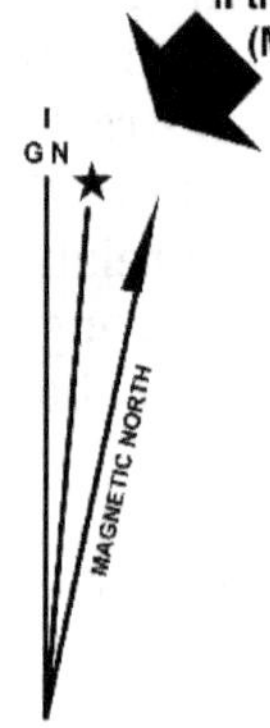

To put your compass reading on your map:

You are converting a Magnetic Azimuth to a Grid Azimuth

You will **ADD** the G-M Angle to the Magnetic Azimuth

MAZ to GAZ → Add

To use your map reading on your compass:

You are converting a Grid Azimuth to a Magnetic Azimuth

You will **SUBTRACT** the G-M Angle to the Grid Azimuth

GAZ to MAZ → Subtract

If the Declination Diagram Looks Like This (Mag North is to the West of Grid North)

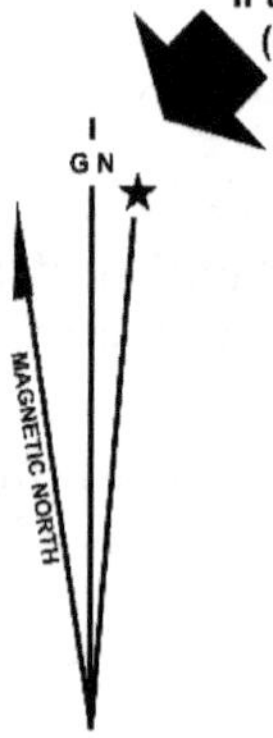

To put your compass reading on your map:

You are converting a Magnetic Azimuth to a Grid Azimuth

You will **SUBTRACT** the G-M Angle to the Magnetic Azimuth

MAZ to GAZ → Subtract

To use your map reading on your compass:

You are converting a Grid Azimuth to a Magnetic Azimuth

You will **ADD** the G-M Angle to the Grid Azimuth

GAZ to MAZ → Add

Moving by Dead Reckoning. Dead reckoning consists of two fundamental steps.

The first is the use of a protractor and graphic scales to determine the direction and distance from one point to another on a map.

The second step is the use of a compass and a means of measuring distance to apply this information on the ground. In other words, it begins with the determination of a polar coordinate on a map and ends with the act of finding it on the ground.

(1) Establish your location by measuring or estimating the distance traveled along the azimuth being followed from the previous known point.

(2) Most dead reckoned movements do not consist of single straight-line distances because you cannot ignore the tactical and navigational aspects of the terrain, enemy situation, natural and man-made obstacles, time, and safety factors. Another reason most dead reckoning movements are not single straight-line distances is because compasses and pace-counts are imprecise measures. Error from them compounds over distance; therefore, you could soon be far afield from your intended route even if you performed the procedures correctly. The only way to counteract this is to reconfirm your location by terrain association or resection (covered in CM-5). Routes planned for dead reckoning generally consist of a series of straight-line distances between several checkpoints with perhaps some travel running on or parallel to roads or trails.

(3) There are two advantages to dead reckoning. Dead reckoning is easy to teach to your team or family and to learn. It is also a highly accurate way of moving from one point to another if done carefully over short distances, even with few external cues are present to guide the movement.

(4) During daylight, across open country, along a specified magnetic azimuth, never walk with the compass in the open position and in front of you. Because the compass will not stay steady or level, it does not give an accurate reading when held or used this way. Begin at the start point and face with the compass in the proper direction, then sight in on a landmark that is located on the correct azimuth to be followed. Close the compass and proceed to that landmark. Repeat the process as many times as necessary to complete the straight-line segment of the route.

(5) The landmarks selected for this purpose are called *steering marks*, and their selection is crucial to success in dead reckoning. Steering marks should never be determined from a map study. They are selected as the march progresses and are commonly on or near the highest points that you can see along the azimuth line that you are following when they are selected. They may be uniquely shaped trees, rocks, hilltops, posts, towers, and buildings—anything that can be easily identified. If you do not see a good steering mark to the front, you might use a back azimuth to some feature behind you until a good steering mark appears out in front. Characteristics of a good steering mark are:

(a) It must have some characteristics about it, such as color, shade of color, size, or shape (preferably all four), that will assure you that it will continue to be recognized as you approach it.

(b) If several easily distinguished objects appear along your line of march, the best steering mark is the most distant. If you have many options, select the highest object. A higher mark is not as easily lost to sight as is a lower mark that blends into the background as you approach. A steering mark should always be visible as you move toward it.

(c) Steering marks selected at night must have even more unique shapes than those selected during daylight. As darkness approaches, colors disappear and objects appear as black or gray silhouettes. Instead of seeing shapes, you begin to see only the general outlines that may appear to change as you move and see the objects from slightly different angles.

(6) Dead reckoning without natural steering marks is used when the area through which you are traveling is devoid of features, or when visibility is poor. At night, it may be necessary to send a member of the unit out in front of your position to create your own steering mark in order to proceed. His position should be as far out as possible to reduce the number of chances for error as you move. Arm-and-hand signals or a radio may be used in placing him on the correct azimuth (use caution). After he has been properly located, move forward to his position and repeat the process until some steering marks can be identified or until you reach your objective.

(7) When handling obstacles/detours on the route, follow these guidelines:
(a) When an obstacle forces you to leave your original line of march and take up a parallel one, always return to the original line as soon as the terrain or situation permits.
(b) To turn clockwise (right) 90 degrees, you must add 90 degrees to your original azimuth. To turn counterclockwise (left) 90 degrees from your current direction, you must subtract 90 degrees from your present azimuth.
(c) When making a detour, be certain that only paces taken toward the final destination are counted as part of your forward progress. They should not be confused with the local pacing that takes place perpendicular to the route in order to avoid the problem area and in returning to the original line of march after the obstacle has been passed.

(8) Sometimes a steering mark on your azimuth of travel can be seen across a swamp or some other obstacle to which you can simply walk out around. Dead reckoning can then begin at that point. If there is no obvious steering

mark to be seen across the obstacle, perhaps one can be located to the rear. Compute a back azimuth to this point and later sight back to it once the obstacle has been passed in order to get back on track.

(9) You can use the deliberate offset technique. Highly accurate distance estimates and precision compass work may not be required if the destination or an intermediate checkpoint is located on or near a large linear feature that runs nearly perpendicular to your direction of travel. Examples include roads or highways, railroads, power transmission lines, ridges, or streams. In these cases, you should apply a deliberate error (offset) of about 10 degrees to the azimuth you planned to follow and then move, using the lensatic compass as a guide, in that direction until you encounter the linear feature. You will know exactly which way to turn (left or right) to find your destination or checkpoint, depending upon which way you planned your deliberate offset.

(10) Because no one can move along a given azimuth with absolute precision, it is better to plan a few extra steps than to begin an aimless search for the objective once you reach the linear feature. If you introduce your own mistake, you will certainly know how to correct it. This method will also cope with minor compass errors and the slight variations that always occur in the earth's magnetic field.

(11) There are disadvantages to dead reckoning. The farther you travel by dead reckoning without confirming your position in relation to the terrain and other features, the more errors you will accumulate in your movements. Therefore, you should confirm and correct your estimated position whenever you encounter a known feature on the ground that is also on the map. Periodically, you should accomplish a resection triangulation using two or more known points to pinpoint and correct your position on the map. Pace counts or any type of distance

measurement should begin anew each time your position is confirmed on the map.

(a) It is dangerous to select a single steering mark, such as a distant mountaintop, and then move blindly toward it. What will you do if you must suddenly call for support or a medical evacuation? You must periodically use terrain association techniques to pinpoint your location along the way.
(b) Steering marks can be farther apart in open country, thereby making navigation more accurate. In areas of dense vegetation, however, where there is little relief, during darkness, or in fog, your steering marks must be close together. This, of course, introduces more chance for error.

(c) Finally, dead reckoning is time-consuming and demands constant attention to the compass. Errors accumulate easily and quickly. Every fold in the ground and detours as small as a single tree or boulder also complicate the measurement of distance.

Some Compass Types and Characteristics

Lensatic compasses. These compasses are designed for military use and feature a sighting mechanism to improve accuracy. They consist of a rotating bezel, a magnetic needle, and a sighting mechanism that allows the user to precisely align the compass with a target. Get yourself a military grade (Cammenga brand) lensatic compass and learn how to use it. This is one place where we will recommend a specific brand. Tritium and phosphorescent each have their advantages, our preference is always Tritium. The marking indicators will fade as the radioactive material becomes dim over time so have a plan to replace a tritium compass every 8-10 years.

Baseplate compasses. These compasses are designed for hiking and outdoor activities. They feature a flat baseplate with a rotating bezel, a magnetic needle, and a ruler for measuring distances on maps. The baseplate is usually a fast settling compass that is outstanding for orienteering and terrain association. They are not the optimal tool for dead reckoning or obtaining an accurate azimuth compared to a lensatic. Quality baseplates are usually liquid filled; when inspecting ensure there are no bubbles present in the sight glass as these can disrupt the needle float. Even a small bubble can throw a bearing off as the needle settles.

Examples of baseplate compasses alongside a lensatic compass

GPS compasses. These compasses use GPS technology to determine direction and location. They are useful in areas where traditional compasses are not effective, such as in heavy urban areas or in the presence of strong magnetic fields.

Lensatic Compass Components

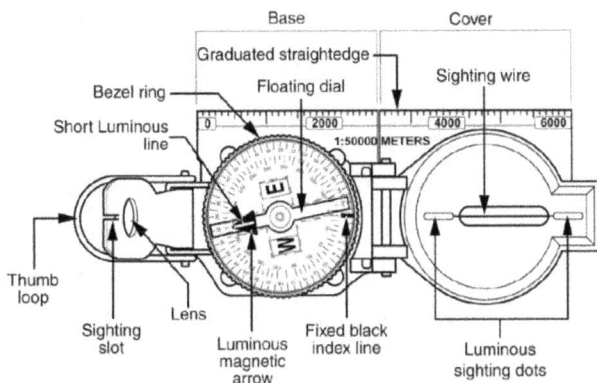

Part	Function
Thumb loop	• Serves as a retaining device to secure the compass in the closed position • Also used as a wire loop for your thumb when you hold the compass in position for sighting in on objects
Cover	• When closed, protects the face of the crystal • Contains the sighting wire
Sighting wire	• Used for: ○ sighting in on objects for which an exact azimuth is needed ○ compass calibration • Is comparable to the front sight post of the service rifle
Bezel ring	• Holds the upper glass crystal in place • Helps preset a direction for night compass navigation • Contains 120 clicks when rotated fully; each click equals 3° • A short luminous line is used in conjunction with the north-seeking arrow for night navigation
Black index line	• Stationary line used as a reference line for determining direction • When the compass is held properly, the azimuth found directly under the black index line identifies the direction that the compass is pointing
Compass dial	• Is delicately balanced and free floating when in use • Can be locked in place by closing the eyepiece • Contains two complete circular scales, one in ○ Degrees (red scale) ○ Mils (black scale)
Lanyard (Dummy cord)	• Helps prevent loss of the compass • Periodically, the brass clamp and the cord itself need to be checked for serviceability

General Inspection. Compasses are precision instruments and should be cared for accordingly. A detailed inspection is required when first obtaining and using a compass. Important serviceability checks are outlined on the next page.

Visual Inspection Checklist.

Cover - Make sure the cover is securely attached and is not damaged or cracked. The cover protects the compass from damage and also serves as a sighting device.

Sight - Check that the sighting mechanism, such as the sighting wire or the sighting slot, is working properly and is not damaged.

Lens - The lens should be free of cracks, scratches, or other damage that could affect the accuracy of the compass readings.

Bezel - The rotating bezel should rotate smoothly and easily. The markings on the bezel should be clearly visible and not worn away.

Needle - The magnetic needle should be free of rust or corrosion and should move easily in response to magnetic fields. The north-seeking end of the needle should be properly aligned with the magnetic meridian.

Dial - The dial should be clearly marked and easy to read. The needle should rotate smoothly and easily over the dial without any slipping or jumping.

Housing - The housing of the compass should be sturdy and free of cracks or other damage that could affect the accuracy of the compass readings.

Shooting an Azimuth

Shooting an azimuth is the task of aiming your compass at a distant object or aligning the compass with a specific degree reading. Magnetic azimuths are determined with the use of magnetic instruments, such as lensatic and baseplate compasses.

Using the Centerhold Technique. First, open the compass to its fullest so that the cover forms a straightedge with the base. Move the lens (rear sight) to the rearmost position, allowing the dial to float freely. Next, place your thumb through the thumb loop, form a steady base with your third and fourth fingers, and extend your index finger along the side of the compass. Place the thumb of the other hand between the lens (rear sight) and the bezel ring; extend the index finger along the remaining side of the compass, and the remaining fingers around the fingers of the other hand. Pull your elbows firmly into your sides; this will place the compass between your chin and your belt. To measure an azimuth, simply turn your entire body toward the object, pointing the compass cover directly at the object. Once you are pointing at the object, look down and read the azimuth (degrees are RED, always Read RED) from beneath the fixed black index line.

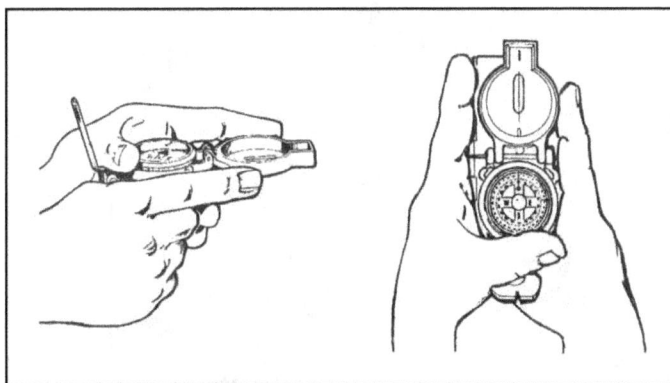

Centerhold Technique

The centerhold method offers the following advantages over the sighting technique:
(1) It is faster and easier to use.
(2) It can be used under all conditions of visibility.
(3) It can be used when navigating over any type of terrain.
(4) It can be used without putting down the rifle; however, the rifle must be slung well back over either shoulder.
(5) It can be used without removing eyeglasses.

Using the Compass-to-Cheek Technique. Fold the cover of the compass containing the sighting wire to a vertical position; then fold the rear sight slightly forward. Look through the rear-sight slot and align the front-sight hairline with the desired object in the distance. Then glance down at the dial through the eye lens to read the azimuth (Read RED). The compass-to-cheek technique is used almost exclusively for sighting, and it is the best technique for this purpose.

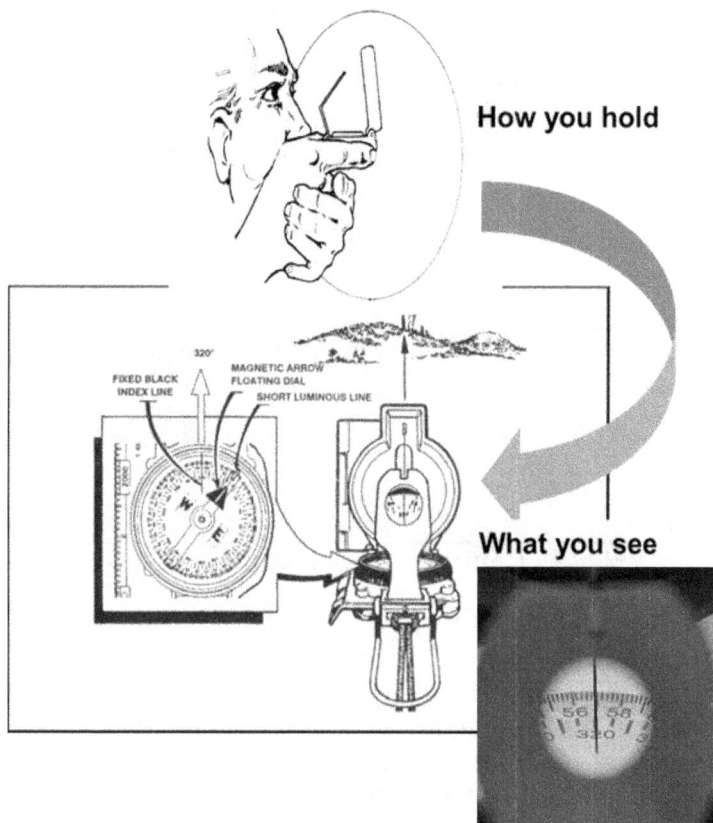

Compass to cheek method

How you hold

What you see

Presetting a Compass and Following an Azimuth

Although different models of the lensatic compass vary somewhat in the details of their use, the principles are the same.

Setting up an Azimuth for Daytime (or with a light source at night):

Hold the compass level in and rotate it until the azimuth falls under the fixed black index line (for example, 320°).

Turn the bezel ring until the luminous line is aligned with the north-seeking arrow to preset the azimuth.

To follow an azimuth, use the centerhold technique and turn your body until the north-seeking arrow is aligned with the luminous line. Move in the direction of the front cover's sighting wire, which is aligned with the fixed black index line that contains the desired azimuth.

Setting up an Azimuth for Night (without a light source):

The azimuth may be set on the compass by the click method. Remember that the bezel ring contains 3° intervals (clicks).

(a) Rotate the bezel ring until the luminous line is over the fixed black index line.

(b) Find the desired azimuth and divide it by three. The result is the number of clicks that you have to rotate the bezel ring.

(c) Count the desired number of clicks. If the desired azimuth is smaller than 180°, the number of clicks on the bezel ring should be counted in a counterclockwise direction. For example, the desired azimuth is 51°. Desired azimuth is 51° ÷ 3 = 17 clicks counterclockwise. If the desired azimuth is larger than 180°, subtract the number of degrees from 360° and divide by 3 to obtain the number of clicks. Count them in a clockwise direction. For example, the desired azimuth is 330°; 360°-330° = 30 ÷3 = 10 clicks clockwise.

(d) With the compass preset as described above, assume a centerhold technique and rotate your body until the north-seeking arrow is aligned with the luminous line on the bezel. Proceed forward in the direction of the front cover's luminous dots, which are aligned with the fixed black index line containing the azimuth.

(e) When the compass is to be used in darkness, an initial azimuth should be set while light is still available, if possible. With the initial azimuth as a base, any other azimuth that is a multiple of three can be established through the use of the clicking feature of the bezel ring.

Sometimes the desired azimuth is not exactly divisible by three, causing an option of rounding up or rounding down. If the azimuth is rounded up, this causes an increase in the value of the azimuth, and the object is to be found on the left. If the azimuth is rounded down, this causes a decrease in the value of the azimuth.

Determining Pace Count

A way to measure ground distance is the pace count. A pace for land navigation purposes is equal to two natural steps, counted on every other foot. Always step off each time with your left foot and count only the left foot strikes. Your 100 meter pace count is probably going to be somewhere in the low 60s to low 70s. To accurately use the pace count method, you must know how many paces it takes you to walk 100 meters. To determine this, you must walk an accurately measured course (GPS or rangefinder) and count the number of paces you take. A pace course can be as short as 100 meters or as long as 600 meters. The pace courses, regardless of length, must be on varied terrain to that you will be walking over. It does no good to walk a course on flat terrain and then try to use that pace count on hilly terrain. Day, night, load variance, terrain type, soil type (sand will significantly add paces) and fatigue level will all influence your pace count. Experiment during training to determine what works for you and what is the most accurate.

Keeping Pace Count. Maintaining count during a movement is critical, do not rely on your memory (was that 5 or 6 kilometers, was that 7 or 800 meters?). Losing count is a real thing under stress, fatigue, and mission focus, these can really disrupt one's ability to keep track of the count. Using pace beads, moving pebbles from one pocket to another, or simply making tick marks in your notebook are all ways to keep track with pace beads being the most efficient and the least distracting.

> *TTP: when planning your route try to shorten the legs to around 500 meters to an identifiable (tactically sound) feature. Shorter legs are easier to manage, keep track of, and will enhance confidence in the route.*

Individual Route Selection and Planning

These principles apply to moving as an individual, two or three man teams, or your movement in tactical formations. OAKOC always applies, and as you develop your skills you will be able to visualize the route over the terrain. It is an amazing blend of art and science, you will begin to mentally traverse the route based on the topo map "lets see...contour lines close together, will be steeper and slow movement but the underbrush will be less and it will most likely be dry; Ill follow the ridgeline north about a third of the way parallel to the ridge to avoid the low ground" etc etc.

Use Terrain for Protection. Terrain offers cover and concealment from observation and fires. Dismounted movement techniques can help units use the terrain over which they move to their advantage. Avoid "skylining" or moving near the top of a ridge or hill while silhouetting yourself. Do your best not stand up and move directly forward from a defilade / hasty overwatch position; move to the flanks behind what you were using as cover and then move forward. An enemy observer that was watching may not have had enough of a target to engage until you stood up from the defilade and moved straight forward (also covered in the IMT section later in this manual). Choose routes that do not stand out as "paths of least resistance". Humans are lazy and will follow these *natural lines of drift* through terrain (the easier path, think of your local park and where the worn human trails are). Combatants are no different; the undisciplined and/or untrained will use trails, choose easier routes, and will avoid swamps and low ground.

Avoid Possible Kill Zones. Avoid large open areas surrounded by cover and concealment or those dominated by a piece of terrain that the enemy would likely use. Watch for the presence of obstacles or any other signs of an enemy engagement area (killing ground that an enemy will likely want an opposing unit to be canalized into).

Move During Limited Visibility. Movement during darkness or other limited visibility conditions provides concealment from enemy gunners at long range. Combining a route's characteristics with the advantages of poor weather or limited visibility advantages during night can work in your favor.

Legs. The best way to manage a route is to divide it into segments called "legs." By breaking the overall route into several smaller segments, you are able to manage the longer route during your movement. Legs typically have only one distance and direction. A change in direction will end the leg and begin a new one.

A leg must have a defined beginning and ending, marked with a graphic control measure such as a checkpoint or phase line. (When using GPS, these are captured as waypoints.) When possible, the start point and end point should correspond to a recognizable navigational aid (catching feature or navigational attack point).

To develop a leg first determine the type of navigation and route best suiting the situation. Once these two decisions are made, select the distance and direction from the start point to the end point and identify critical METT-TC information as it relates to the specific leg. If time allows and requirements dictate doing so you can capture this information and draw a sketch on a route chart (an example is in the diagram below). Do not take this on the mission, this is simply a visualization, planning, and rehearsal tool for small unit leaders to use.

LEG	AZIMUTH / DISTANCE	KEY INFORMATION	
Leg 1: SP 1 to CKP 1. - Stay in woodline east of the hardball at the base of the hill.	008°/500m	O: Limited. A: HWY 1. K: Hill mass west of leg. O: N/A. C: Good.	
Leg 2: CKP 1 to CKP 2. - Stay on south side of dirt secondary road. Continue movement to the large church at CKP 2.	030°/1800m	O: Unlimited. A: Dirt trail. K: Hill 18 southeast of leg. O: N/A. C: None.	
Leg 3: CKP 2 to RP. - Stay on east side of hill 25. Continue movement to the boulders at the RP.	319°/450m	O: Unlimited. A: Dirt trail. K: Hill 25 west of leg. O: N/A. C: None.	

LEGEND			
CKP	CHECK POINT	SP	START POINT
RP	RELEASE POINT	M	METERS

Bypassing an Obstacle. To bypass enemy positions or obstacles and still stay oriented, detour around the obstacle by moving at right angles for specified distances.
(1) For example, while moving on an azimuth of 90° change your azimuth to 180° and travel for 100 meters. Change your azimuth to 90°and travel for 150 meters. Change your azimuth to 360°and travel for 100 meters. Then, change your azimuth to 90°and you are back on your original azimuth line.

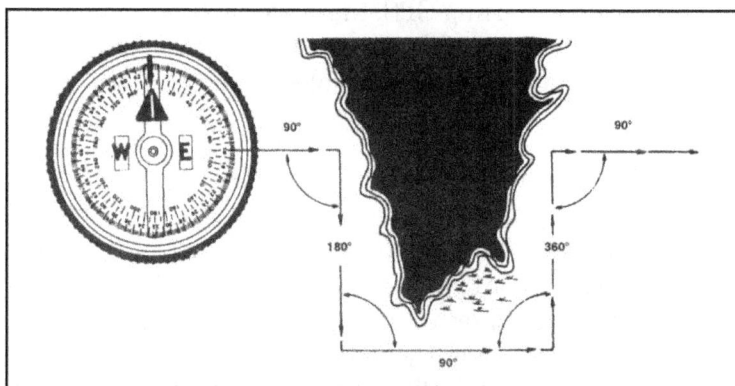

(2) Bypassing an unexpected obstacle at night is a fairly simple matter. To make a 90° turn to the right, hold the compass in the centerhold technique; turn until the center of the luminous letter E is under the luminous line (*do not* move the bezel ring). To make a 90° turn to the left, turn until the center of the luminous letter W is under the luminous line. This does not require changing the compass setting (bezel ring), and it ensures accurate 90° turns.

Offset. A deliberate offset is a planned magnetic deviation to the right or left of an azimuth to an objective. Use it when the objective is located along or in the vicinity of a linear feature such as a road or stream. Because of errors in the compass or in map reading, the linear feature may be reached without knowing whether the objective lies to the right or left. A deliberate offset by a known number of

123

degrees in a known direction compensates for possible errors and ensures that upon reaching the linear feature, the user knows whether to go right or left to reach the objective. Ten degrees is an adequate offset for most tactical uses. Each degree offset moves the course about 18 meters to the right or left for each 1,000 meters traveled. For example, in the figure below, the number of degrees offset is 10. If the distance traveled to "x" in 1,000 meters, then "x" is located about 180 meters to the right of the objective.

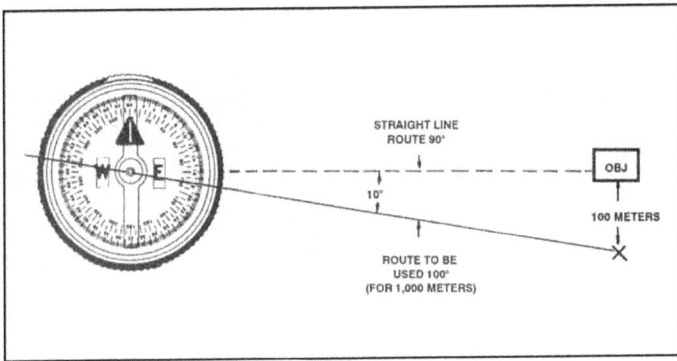

Backstop. A backstop is an easily identifiable terrain or man made feature that swill indicate you have gone too far beyond your destination or checkpoint. Roads and streams make great backstops, these are the "when I reach the road I know I have gone too far" measure.

Attack Point. An easily recognizable feature that is a few hundred meters from your end point / objective that you can easily find. You will use the attack point to "reset" your nav route, they can assist with finding more difficult objectives that may not be near an identifiable feature. Once you reach your attack point double check your location and then execute the final short leg you had already planned. This shorter (and more manageable) leg will bring you to your final objective.

Handrail. A handrail is similar to the backstop but they run parallel or partially parallel to your desired route and will indicate departure from your route laterally. These can also be used (if the tactical situation dictates) to guide navigation by walking to the side but within visual of the feature. Again linear features like roads and streams work best, just use caution with the natural lines of drift and assess the risk using the METT-TC and OAKOC tools.

Navigation roles you may fill in a tactical formation.

Compass Man. The compass man assists in navigation by ensuring the lead fire team leader remains on course at all times. The compass man should be thoroughly briefed. His instructions must include an initial azimuth with subsequent azimuths provided as necessary. The platoon leader or squad leader also should designate an alternate compass man. The leader should validate the patrol's navigation with GPS devices if available.

Pace Man. The pace man maintains an accurate pace at all times. The platoon leader or squad leader should designate how often the pace man reports the pace. The pace man also should report the pace at the end of each leg. The platoon leader or squad leader should designate an alternate pace man.

Tactical Movement

Tactical movement involves movement of a unit assigned a mission under combat conditions when not in direct ground contact with the enemy. Tactical movement is based on the anticipation of early ground contact with the enemy, either enroute or shortly after arrival at the destination. Movement ends when ground contact is made or the unit reaches its destination. *Movement is not maneuver.* Maneuver is conducted while in contact, supported by fire, to gain a position of advantage over the enemy.

There is considerable overlap between the two, and units transition from one to the other during actions on contact. You must reduce your exposure to the enemy during movement by use of the terrain, avoidance of possible kill zones, dispersion, reconnaissance, and the use of measures to counter enemy observation and fires.

The key to movement involves selecting the best combination of combat formation and movement technique for each situation. Leaders consider METT-TC in selecting the best route and appropriate formation and movement technique. The leader's selection must allow the moving unit to:

Maintain cohesion.

Maintain communication.

Maintain momentum.

Provide maximum protection.

Make enemy contact in a manner allowing them to transition smoothly to offensive or defensive action.

Moving in Tactical Formations

Before we discuss formations and movement techniques we will address the fundamentals of moving as a member of a team (any size) in a tactical formation. Remember stealth and remaining undetected are always the requirement for an irregular force. Training and getting reps in with your team during both day and night will build group and individual proficiency through repetition and familiarity. You will be able to recognize your teammates simply by the way they walk at night, you will know how they move, and will recognize what small gestures and movements indicate. Moving in a formation is an exercise in trust and familiarity. When the leader or the man in front of you stops, takes a knee or goes prone so will you. You will trust that he is doing what he does for a reason. Good units will execute change of formation drills when directed, great units will change formations automatically (based on SOP) as terrain and conditions change, they will flow smoothly like water over the terrain.

Even well-camouflaged fighters give off indicators of presence. Movement, shapes, shine are all liabilities.

SLLS

The SLLS (pronounced "sills") technique of **S**top, **L**ook, **L**isten, **S**mell (aka Listening Halt) is a method used to adapt senses to the sights, sounds, and smells of the battlefield. Allowing your senses to adapt to the environment will increase your sensitivity to subtle changes that may indicate a threat. Birds chirping that suddenly stop, a faint smell of wood smoke, or a distant horizontal shape deeper in the woodline may have gone unnoticed if your unit had just continued directly into the tactical movement. SLLS halts should be done shortly after departure for a mission in a covered and concealed position and repeated as the leader assesses the need to do so or if there is a significant change in conditions.

The SLLS process also applies to our own unit, just as we assess terrain we need to consider SLLS elements that you can address before we go out on mission during Pre Combat Inspections (PCI). Showering with highly scented soap, shaving cream, clean laundry, uncamouflaged weapons...these all are signs of enemy presence that you will be looking for during SLLS halts. Ensure you are not giving off the very indicators that you and your team are looking for in the threat. Remember SLLS principles apply to both friendly and enemy; assess yourself by visualizing your own unit's current state from the enemy perspective.

Stop. This step involves stopping in place to assess the situation. It is important to take a moment to focus on your surroundings and gather information about what is happening in the area and let your senses adjust.

Look. This step involves using your eyes to observe and gather information about the environment. Look for signs of movement, changes in the environment, or anything that seems out of place. Visual indicators include shape, shadow, shine, silhouette, surface, and movement.

128

Listen. Listen for sounds that may indicate the presence of danger or provide information about the environment. Pay attention to the sounds of animals, vehicles, or people and also the *absence* of sounds that may be an indicator.

Smell. Use your sense of smell to detect potential dangers or gather information about the environment. Be aware of unusual smells, such as wood or cigarette smoke, vehicle exhaust, food preparation or chemicals that may indicate the presence of a threat. The longer you remain in the field without access to showers and civilization the more sensitive you will be to unnatural scents. A freshly showered individual, weapons cleaning chemicals, or even slight food smells become very noticeable over distance after spending a few days in the field.

Your basic responsibilities in a formation

Noise Discipline. Having gear squared away and silenced is only part of the equation. Being able to take steps and walk quietly through the woods and silently communicate with teammates using hand and arm signals (both standard and unit specific) is imperative. Sound always travels further than you think, which means that slow, deliberate movement will be to your advantage. Accessing gear during halts must be done silently, practice and rehearse so you know what you typically need to access during a tactical movement and adjust your load setup accordingly (eg don't keep the gear you habitually need to access in a velcro enclosed pouch; opening in the quiet woods is a non-starter for noise discipline). Radios must be managed to prevent radio traffic from making noise through an external mic or speaker. Slow, methodical footfall by rolling heel to toe or quietly placing your feet down is a woodsman and soldier skill you need to develop. You will be making long dismounted movements so this quiet walking requirement is something that you will have to employ long term, under load, and while physically exhausted. Overloaded troopers can move quietly for a very limited amount of time, noise discipline falls apart quickly for physically fatigued formations. Practical experience is the only way to gain an appreciation for how quiet or loud a group of fighters is during a long duration mission.

Always know where you are. In uncertain situations you as a team member may be called upon to perform leader duties, call for assistance, or you may simply be separated and need to make it back to the last rally point (RP). Having a general idea of where you are at all times will allow you to shorten the time it takes to reorient and take action. Cruising along fat, dumb, and happy is a recipe for getting lost, separated, or wandering into a known enemy location.

Always know where your battle buddy is. We always use the buddy system. The two of you will go everywhere together. If there is an odd number, then there will be three man buddy team. Accountability and just simply looking out for each other must be drilled into your head as normal. Two man recon teams, buddy team rushes (more on this later) and security rotations are all built on this concept.

Know the Challenge and Password. We use a challenge and password system among friendly units that allows us to conduct an initial verbal IFF (Identification Friend or Foe) when in the field or protecting a fixed site. The challenge and password will change at least daily and must be distributed down to each member of the team. You must know the current challenge and password for the present time period and the new ones as they change. You will also have a running password / number combination inside your unit. This is not centrally established or distributed, it is one you and your small unit use in case you are running or being chased back toward your teammates and cannot be easily identified.

Dispersion. Dispersion between individuals and units prevents units from becoming fixed by one enemy position or weapons systems. It also prevents enemy artillery or mortar fires from suppressing the entire element. During movement with teammates it is a balance of maintaining dispersion while keeping the ability to see and support your teammates. Controlling your interval limits the lethal consequences of ambush from either small arms or explosives such as IED's, mortar/ artillery strikes and grenades. Spacing will be adapted to terrain and weather conditions or as directed by the leader. For example, in dense summer woodlands spacing may have to be reduced to maintain visual contact with team members, while in flat open areas such as desert or grasslands your formation will open up significantly. The general rule is that spacing should be kept as wide as possible while maintaining visual contact with your team.

Scanning / Observation. Your unit or group will have SOPs but small unit leaders can also assign responsibilities of observation to individuals in a formation. Scan your assigned sectors enroute while maintaining contact with your team members to reduce the chances of the enemy surprising your unit. It is important to remember not to simply walk along with your eyes fixed on your teammate in front of you. Zoning out (especially when fatigued or at night during a long movement) or getting **tunnel vision** is easy to do especially when not in a leadership position or performing an active navigation role (pace, compass etc). Exhaustion will cause us all to switch off our brains while our feet are still moving along. We need to be self-aware of this and fight through the inclination to just follow along in the woods. You should be in a constant cycle of scanning the surrounding area for potential threats, checking to the front, flanks and rear for teammate location and actions.

Light Discipline. Light discipline is simply common sense, it is just a matter of being aware of the sources and the risks for negligent light discharges. Light discipline involves eliminating or minimizing how much light is emitted from an individual, unit, patrol, or even a fixed site such as your neighborhood

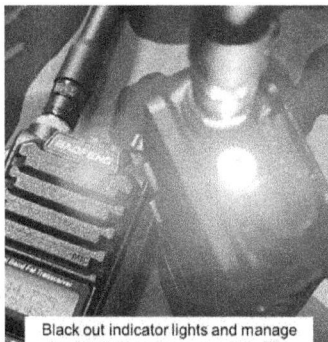

Black out indicator lights and manage backlit displays on your radios. The "flashlight" on a UV5R can have a mind of its own and get turned on by accident.

traffic control points or your shelter (home). This manual is focused on light discipline in the field, but the same concepts apply during any emergency event in the neighborhood. Electricity indicates that you have resources, lights are a civilization beacon and become a great point for an enemy raider to be drawn to at night. Minimizing light output is everyone's responsibility, ensure family members are fully briefed and trained. On patrol or in the field sources of light may be from weapon mounted lights (WML), night vision devices (spill from the rear of the unit), handheld lights, and electronic devices such as radios, watches, weapon mounted lasers, and GPS units. Covering WML lenses when not in use to prevent light negligent discharges and reflection is a best practice, there are several great devices available on the market or you can use a simple homemade cover (test under night conditions to make sure it works). Tail switches should be positioned as to not be bumped by mistake, rail switches should be out of the way when you don't need them but available and repeatable when you do. Red lens lights were the standard for decades, however with near peer threat access to night vision devices using a red lens in the open to do a map check is a high risk action. Even the subdued red (or any other color) lens can be spotted at great distance under NVGs.

A few common sense best practices we recommend including these in unit SOPs and weekend training:

No Smoking – a cigarette can be seen (and smelled) at great distance under the right conditions.

No Campfires – Fires can be seen for miles at night, and in as we apply SLLS tenets to ourselves we understand that it is a huge indicator for smell as well.

Flashlight Management – High candela weapon and handheld lights should only be used at specific times. Once you hit that tail switch it is akin to sounding an alarm. Even a short blip of white light can be seen at distance. Small red lens low lumen lights are ok for close use under cover of a poncho or similar, sometimes there is no way around using one. These principles apply to IR sources as well, a full power IR laser that is negligently discharged can be just as risky if the threat has any kind of passive NV. There are recent examples from the Ukranian war of IR lasers being inadvertently left in "constant on" mode in defensive positions acting as an accidental beacon.

Cover reflective surfaces – camouflage or cover reflective surfaces (day and night). Mute or dull the reflection or IR glow from gear and weapons. Be aware of your optics and glint during the day, glass can reflect and act as a signal mirror even on overcast days.

Indicator Lights - completely blackout or disable sensor or indicator lamps on electronics. Transmit/receive, the "flashlight" on the Baofeng radios, and even the heartrate sensor on some smart watches are sources of light that must be eliminated.

Signal Discipline. Use radios only when necessary, remember anything you say on the radio waves can be heard by others. The signal can not only be detected, it can also indicate your location by direction finding equipment. Using radio listening silence (RLS) to keep your team members off the radio net until absolutely necessary is always a good practice. Signal emitters that push an electronic signal out without the direction of the end user must be left behind or disabled. Wifi, cellular, bluetooth signals are all detectable albeit very short range in most cases. Mobile phones are a no go, these are easily tracked and will compromise a mission. The ability of a third party entity to turn a device on, use features remotely or pull meta data after the fact is real. Leave them behind, conditions may be such that destruction of all phones may be necessary. Small devices such as wireless mics, smart watches, and even some cameras and rangefinders have bluetooth that broadcast short range signals that can be a risk to the mission. The unmanned aerial systems operators on your team have got to be absolute experts at managing this vulnerability. Drones are an incredible asset for the Pro Citizen, but in near peer or superior force environments the electronic trail that unmanned systems emit can be leveraged against you. Weighing the risk, cueing the assets based on the electronic detection threat, and good use of terrain masking are just the starting point for drone use consideration. With any modern organized force as an opponent the use of UAS will most likely be a no go. Situational awareness and C2 (Command and Control) systems are extremely high risk for use. Targeting data is a foregone conclusion should you choose to use any of these commercial systems.

Litter Discipline. Leave no trace. Discarded wrappers, food, lost pieces of gear or equipment, or spent batteries are all indicators of presence in an area. You can't do much about sign left behind such as tracks and vegetation disturbance but policing up after yourself goes a long way toward hiding presence in an area. When prepping for combat tie your gear down to prevent field loss. Field strip MREs and other food items to minimize trash generation during a patrol. Include small resealable trash bags in your loadout for longer duration missions. Any trash should be packed out, buried trash can still be found by an enemy and exploited for pieces of intel. Burying some things can't be avoided, but you should minimize the practice.

Keep Your Camouflage Fresh. As time wears on during a patrol individuals can get lazy or sloppy with their camo. Keep yourself and your battle buddy squared away to help your patrol remain in the shadows. Camo makeup wears off quickly (unless you are actually trying to wash it off and get cleaned up, it is one of those tactical paradoxes). Sweat and movement will create bare spots, personal hygiene can remove makeup and create shine. Touch up camo, square gear away, keep lenses covered, and any vegetation you decide to use should be replaced as it wilts or appears out of place. This also applies across seasons, replace summer/green based camo patterns in autumn and winter months.

Terms and Definitions you need to be familiar with that are associated with formation movement, maneuver, and tactical operations in general. Don't get wrapped around the axle on learning these in detail at this point, just being familiar with what they are and why they exist will assist your understanding as you hear these terms during training events and classes with your group.

BMNT - Begin Morning Nautical Twilight. This is the time of morning when in good conditions and in the absence of other illumination, enough light is available to identify the general outlines of ground objects and conduct limited military operations. Light intensification devices are still effective and may have enhanced capabilities. In the iron sight days we called this "first shooting light". It was just enough illumination with the naked eye to make out shapes. Positive ID is not possible under these conditions. BMNT occurs before sunrise, can be affected by weather or smoke, and can be impacted by existing snow cover. Movement planning can be queued off of this time, the flip side is this is when unknown/adversarial individuals (without the aid of night vision) can start effectively moving about. BMNT is also the time that the woods "wake up". Ground animals will start to move about and birds will begin chirping. Presence or absence of these can be an indication of human or predator movement. The US Naval Observatory website has light data tables that you can select by date that will indicate when BMNT occurs.

Stand To - Security stand-to (everyone is awake and ready), security is at 100 percent and all systems are combat ready. Accountability of all personnel by first line leaders is relayed to the leadership prior to stand to. The patrol leader designates the night before when stand- to is the next morning; unit TACSOPs will outline what actions must occur.

REDCON 1 – Readiness Condition 1 is when the element is at full alert, you are ready to move and fight. The thumbs up or radio call stating you are REDCON 1 means you are literally ready to stand up and start movement (when vehicle mounted it means your vehicle and your crew are at their positions and ready to roll). All systems are turned on, PCCs are completed.

PCC – Pre Combat Checks PCCs occur at all levels as an individual task, PCCs must be conducted during Step 4 of TLPs (Start Necessary Movement). Individual team members must be made to understand the significance of the responsibility of ensuring they are mission capable. They must be mentored on the importance of self- and battle-buddy PCCs. Pre-Combat Checks ensures that teammates at all levels are adequately prepared to execute the mission. PCCs are specific mission oriented checks that a unit conducts before operations.

PCI – Pre Combat Inspections PCIs must occur at all leader levels, to include team leaders and squad leaders. PCIs should remain in Step 8 (Supervise and Refine) of the TLPs. PCIs are executed by leaders as a spot check of mission critical items. This includes an individual's knowledge of the upcoming mission (what unit is to our left, what graphic control measure is our limit of advance etc etc). If deficiencies arise from PCIs, the leader may have follow-up PCCs.

WARNO – Warning Order (the current version was changed to "WARNORD", some staff member got an Army achievement medal for changing that one). We do what we want to do, for our purposes it is a WARNO. A WARNO is a time saver to allow parallel planning by subordinates and is a preliminary notice of an order or action to follow. After the leaders receive new missions and assess the time available for planning, preparing, and executing the mission, they immediately issue WARNOs to their subordinates to enable them to begin their own planning and preparation (parallel planning) while they begin to develop the OPORDs. When they obtain more information, they issue updated WARNOs, giving subordinates as much information as they know.

OPORD – Operations Order, or a five paragraph operations order is a specified format order that concisely and clearly issues direction and purpose to a subordinate unit. Unit leaders will receive OPORDs from their immediate headquarters and will analyze them (Mission Analysis). This analysis is essentially translating the higher directives and purpose into actions for the unit. In turn the subordinate leaders will do the same process and issue orders to their subordinates. The OPORD paragraphs are

Situation (terrain and weather, friendly and enemy forces), the *Mission* (the who, what, where, and when), *Execution* (the scheme of maneuver and concept of how the mission will be executed) as well as the commander's intent (the big "why" or intended end state of what we are doing). Commander's intent is the clear, concise statement of what the force must do and the conditions the force must establish with respect to the friendly, enemy, terrain, and civil considerations that represent the commander's desired end state. This serves to allow subordinate and supporting commanders to achieve the commander's desired results without further orders, even when the operation does not unfold as planned. Paragraph 4 is *Sustainment* (addresses supply, medical, and equipment repair actions to name a few) and Paragraph 5 is *Command and Signal* covering communications and leadership locations and succession of command. We will cover this in detail in patrolling and small unit operations.

TLP – Troop Leading Procedures are the 8 steps that leaders use to provide small-unit leaders a framework for planning and preparing for operations. TLP begin when the leader receives the first indication of an upcoming mission and continues throughout the operational process (plan, prepare, execute, and assess). TLPs comprise a sequence of actions helping the leader use available time effectively and efficiently to issue orders and execute operations. They are not a hard and fast set of rules, some actions may be performed simultaneously (actually desired) or in a sequence different than listed in doctrine. They are a guide being applied consistent with the situation and experience of the leader and his subordinate leaders. All steps should be done, even if in abbreviated fashion. As such, the suggested techniques are oriented to help a leader quickly develop a combat order.

Mission Type Orders – mission orders are not a specific type, they describe the "style" or characteristic of permissiveness in that they are designed to enable disciplined initiative within the commander's intent to empower subordinate leaders. In the absence of further orders and changing conditions a subordinate leader can apply judgment and still accomplish the goal of the mission even if he has to improvise a solution. This has been the strength of our military since the 1980s, having a corps of small unit leaders (NCOs and company grade officers) that can be trusted to do what is required and achieve the desired outcome. Applying this characteristic when issuing orders or even giving mundane tasks to subordinates will build trust, confidence, and lead to some outstanding organizations.

GOTWA (you and me) – is a mnemonic that stands in for the Five Point Contingency Plan (basic info given to the next in charge when a leader goes away from the main unit to receive an order, conduct a leader recon etc). It stands for where I am **G**oing (to include routes to and from), **O**thers I am taking with me, **T**ime I will be away/will return, **W**hat to do if I don't come back, and **A**ctions on contact for you (the unit) and me (while I am enroute or away).

FRAGO – Fragmentary Order is issued to implement a change or adjustment to an existing order and may come at any time. It allows leaders flexibility and cuts planning time by allowing them to issue the minimum instructions to adjust from the current order. FRAGOs are typically issued when adjustment decisions, not execution decisions must be made. A FRAGO is either oral or written and addresses only those parts of the original OPORD that have changed. It is issued in the same sequence as the OPORD (many items or even entire paragraphs may simply state "no change").

Basic Dismounted Formations

We will address a few of the basic formations for the fire team and squad so you can familiarize yourself with what they look like (without terrain). Other manuals in this series address unit movement in greater detail, however you need to understand early on where and how you fit into a dismounted tactical formation.

We start with standardized formations to learn to control movement, optimize firepower and security, and ensure combat power is synchronized with each other while executing combat operations. Every individual has a standard position in a formation that allows seamless operations with other friendly forces. Leaders control their units using arm-and hand signals, verbal commands during contact, and intra-squad/team communications (on a very limited basis due to electronic signature).

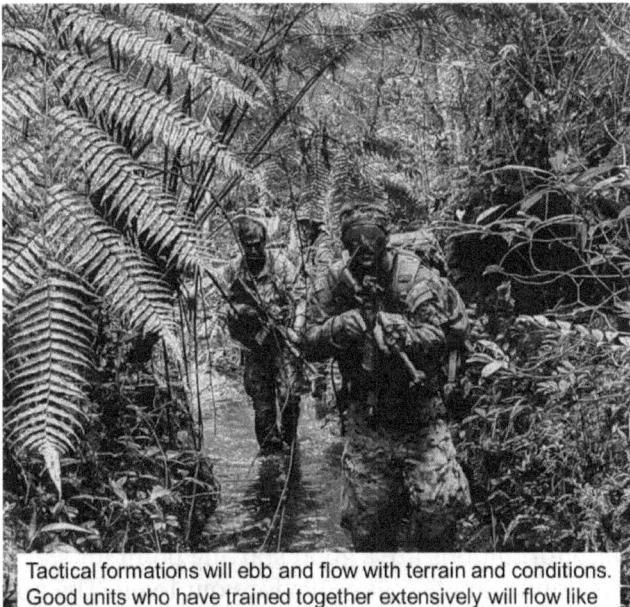

Tactical formations will ebb and flow with terrain and conditions. Good units who have trained together extensively will flow like water over terrain, expanding and contracting as visibility and the situation changes.

Seven Dismounted Formation Examples and the associated hand and arm signal:

NAME / FORMATION / SIGNAL (IF APPLICABLE)	CHARACTERISTICS	ADVANTAGES	DISADVANTAGES
Line Formation	- All elements arranged in a row - Majority of observation and direct fires oriented forward; minimal to the flanks - Each subordinate unit on the line must clear its own path forward - One subordinate designated as base on which the other subordinates cue their movement	Ability to: - Generate fire superiority to the front - Clear a large area - Disperse - Transition to bounding overwatch, base of fire, or assault	- Control difficulty increases during limited visibility and in restrictive or close terrain - Difficult to designate a maneuver element - Vulnerable assailable flanks - Potentially slow - Large signature
Column/File Formation	- One lead element - Majority of observation and direct fires oriented to the flanks; minimal to the front - One route means unit only influenced by obstacles on that one route	- Easiest formation to control (as long as leader can communicate with lead element) - Ability to generate a maneuver element - Secure flanks - Speed	- Reduced ability to achieve fire superiority to the front - Clears a limited area and concentrates the unit - Transitions poorly to bounding overwatch, base of fire, and assault - Column's depth makes it a good target for close air attacks and machine gun beaten zone
Vee Formation	- Two lead elements - Trail elements move between the two lead elements - Used when contact to the front is expected - "Reverse wedge" - Unit required to two lanes/routes forward	The ability to: - Generate fire superiority to the front - Generate a maneuver element - Secure flanks - Disperse - Transition to bounding overwatch, base of fire, or assault	- Control difficulty increases during limited visibility and in restrictive or close terrain - Potentially slow
Box Formation	- Two lead elements - Trail elements follow lead elements - All-around security	Same as vee formation advantages	Same as vee formation disadvantages
Wedge Formation	- One lead element - Trail elements paired off abreast of each other on the flanks - Used when the situation is uncertain	The ability to: - Control, even during limited visibility, in restrictive terrain, or in close terrain - Transition trail elements to base of fire or assault - Secure the front and flanks - Easy transition to line and column	- Trail elements are required to clear their path forward - Frequent need to transition to column in restrictive, close terrain
Diamond Formation	- Similar to the wedge formation - Fourth element follows the lead element	Same as wedge formation advantages	Same as wedge formation disadvantages
Echelon Formation	- Elements deployed diagonally left and right - Observation and fire to both the front and one flank - Each subordinate unit on the line clears its own path forward	- Ability to assign sectors that encompass both the front and flank	- Difficult to maintain proper relationship between subordinates - Vulnerable to the opposite flanks

Standard formations are a good starting point for learning and developing applicable SOPs for your team. The non-standard force will not be a direct lift from these as you and *your unit may move and operate in a different manner.*

However, there are many fundamentals and principles from these that can be applied to your training. The advantages attributed to one of these formations may be disadvantages to another. Knowing the advantages and applications of these is required to adjust for terrain and enemy situation (METT-TC applies as always). Each does so with different degrees of emphasis resulting in unique advantages and disadvantages. Do not be finnicky in the application of these; variations and modifications should be experimented with and proven out during your training. Keep your tactics grounded in fundamental concepts, train, and rehearse so everyone understands what they are doing and more importantly why they are doing it. Apply some common tactical sense and it does not matter if a formation or tactic doesn't exactly match the accepted doctrine. Ideas that are not tactically sound do not fall into this category.

Do what works for you and your group; don't train from a place of fear that a particular method will not be legitimate because it does not look exactly like doctrine.

Formations with more than one lead element are better for achieving fire superiority to the front but are more difficult to control. Conversely, formations with only one lead element are easier to control but are not as useful for achieving fire superiority to the front.

Leaders must maintain flexibility in their formations to enable them to react when unexpected enemy actions occur. We will show the various dismounted formations, however we will focus on the fire team Wedge and File in this manual.

The term fire team formation refers to your relative position within the fire team and how the team is arrayed on the battlefield as they move across various terrain. Fire team formations include the fire team wedge and fire team file. Regardless of which formation the team employs, each person must know his location in the formation relative to the other fire team members and team leader (TL). Each individual covers a set area of responsibility for observation and direct fire as the team is moving. To provide the unit with all-around protection, these areas interlock. Team leaders are constantly aware of their teams' sectors of fire and correct them as required.

The team or squad leader adjusts the team's formation as necessary while the formation is moving. The distance between individuals is determined by the mission, the nature of the threat, the closeness of the terrain, and visibility. As a general rule, the formation should be dispersed up to the limit of control. This allows for a wide area to be covered, makes the patrol's movement difficult to detect, and makes it less vulnerable to enemy attack. Fire teams may act independently, as part of a squad, or on very rare occasions as an irregular component of an even larger unit.

The Point Man. The importance of an outstanding point man cannot be overstated. The point man is the first in front of a column, squad or team who is responsible detecting mines, IEDs, and spotting enemy ambushes. Should visual or direct fire contact occur he will (in most instances) be the first to make a tactical decision. He scans the terrain and ground in front and around him for indicators of threat presence. The point man is not a sacrificial role, that is draftee Vietnam thinking. He is selected to be the expert at stealthy movement and detecting the enemy. His actions can have operational and sometimes strategic consequences. Understanding the commander's intent and having a sense of the overall situation in the AO is critical for a point man, just as much as any of the other small unit leaders. There is more than one historical example of small unit point men mistakenly initiating direct fire contact with regimental size enemy that resulted in multiple day battles. The squad leader will rely on him to apply his guidance and set the tempo for formation movement. The point man will select approaches through vegetation and terrain, will adjust the pace off the formation behind him, and will be (in most cases) the forwardmost eyes and ears of the SL. He has to be a critical thinker, have a cool head, understand mission type orders, understand the commander's intent, and most likely has a background as an avid outdoorsman. Bow hunters and other big game hunters make for great point men so long as they have the other necessary attributes.

A mature, apex predator-like Point Man is worth his weight in gold. He will keep your patrol out of trouble.

Formations
There are two basic formations for the fire team that other formations will be built upon. The *Fire Team Wedge* and the *File*.

Wedge. The wedge is the basic formation of the fire team.

MOVEMENT FORMATION	WHEN MOST OFTEN USED	Movement Characteristics			
		CONTROL	FLEXIBILITY	FIRE CAPABILITIES AND RESTRICTIONS	SECURITY
Fire team wedge	Basic fire team formation	Easy	Good	Allows immediate fires in all directions	All-round

Wedge extend the arms downward and to the sides at a 45-degree angle below the horizontal, palms to the front. Alternately, use the non-firing hand and point the index finger and pinky finger to the ground.

Fire Team Wedge

TL

Direction of Travel

The interval between individuals in the wedge formation is approximately 10 meters but the wedge expands and contracts depending on the terrain. Fire teams modify the wedge when rough terrain, poor visibility, or other factors make control of the wedge difficult. The normal interval is reduced so all team members still can see their team leader and all team leaders still can see their squad leader. The sides of the wedge can contract to the point where the wedge resembles a single file. Team members expand or resume their original positions when moving in less rugged terrain where control is easier.

In the wedge formation the fire team leader is in the lead position with his team members echeloned to the right and left behind him. The positions for all but the leader may vary. You may have suppressive riflemen, designated marksman, machine gunners (things may develop to make this a possibility), unmanned aerial and ground controllers, and snipers inside our non-standard / guerilla formations. Placing these assets inside of formations is the responsibility of the leader, he will adjust assets location based on METT-TC. This fire team wedge formation permits the fire team leader to lead by example; his standing order to his team is always "Follow me and do as I do." When he moves to the left, his team should move to the left. When he fires, his team fires. When using the lead-by-example technique, it is essential for all individuals to maintain visual contact with their leader and cue off of his actions.

MOVEMENT FORMATION	WHEN MOST OFTEN USED	Movement Characteristics			
		CONTROL	FLEXIBILITY	FIRE CAPABILITIES AND RESTRICTIONS	SECURITY
Fire team file	Close terrain, limited visibility, dense vegetation	Easiest	Less flexible than the wedge	Allows immediate fires to the flanks, masks most fires to the rear	Least

File. Team leaders use the file when employing the wedge is impractical. This formation most often is used in severely restrictive terrain or inside buildings/warehouses, during periods of limited visibility etc. The distance between individuals in the column changes due to the situation, particularly when in urban operations.

Fire Team File

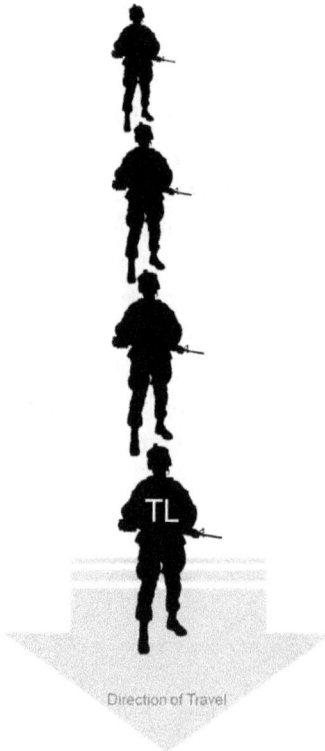

File (Column) Formation - move the non-firing hand to touch the rim of the headgear directly in front of the face. Fingers will be extended and

TL

Direction of Travel

Movement Techniques

Individual movement techniques include high and low crawl, and three to five second rushes from one covered position to another. Individual movement techniques are not prescribed or directed; the individual will decide when to employ each based on the current situation.

Movement techniques for formations are not fixed. They refer to the distances between individuals, teams, and squads vary based on mission, enemy, terrain, visibility, and other factors affecting control. There are three movement techniques: *traveling; traveling overwatch; and bounding overwatch*. The selection of a movement technique is based on the likelihood of enemy contact and the need for speed. Factors to consider for each technique are control, dispersion, speed, and

MOVEMENT TECHNIQUES	WHEN NORMALLY USED	CHARACTERISTICS			
		CONTROL	DISPERSION	SPEED	SECURITY
Traveling	Contact not likely	More	Less	Fastest	Least
Traveling overwatch	Contact possible	Less	More	Slower	More
Bounding overwatch	Contact expected	Most	Most	Slowest	Most

security. As an irregular force Citizens may adapt formations and techniques of movement for use down to two and three man teams. We will discuss this section in terms of squads since it demonstrates the different options well, however you must adapt and adjust this baseline for the team and assets you have available.

Traveling (9 man squad example)

Traveling is **used when contact with the enemy is not likely** and speed is required for the mission. As illustrated in the diagram below the squad leader (SL) is positioned where he can direct both teams with approximately 20 meters (METT-TC dependent) between SL and B TM Leader. When using the traveling technique, all unit elements move continuously. In continuous movement, all individuals travel at a moderate rate of speed, with all personnel alert. During traveling, formations are essentially not altered except for effects of terrain.

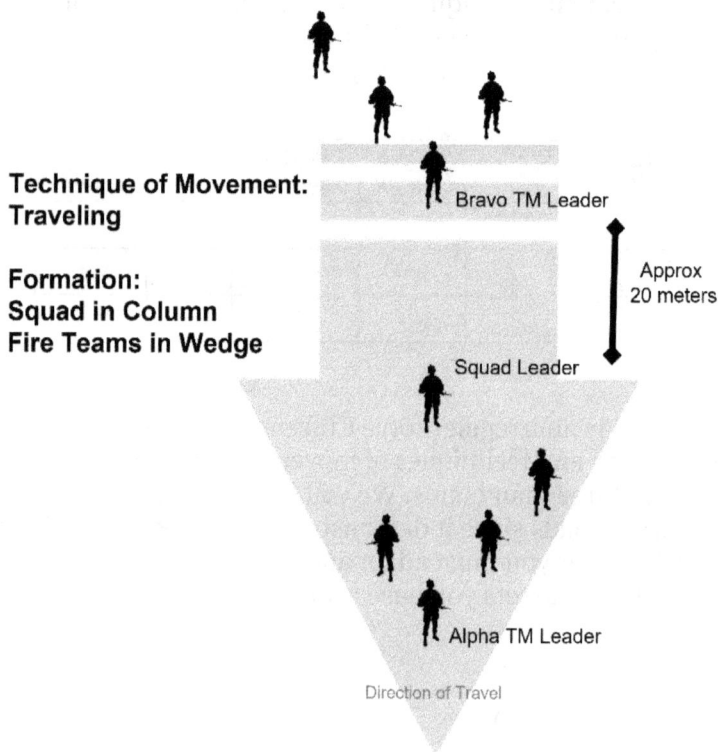

Technique of Movement: Traveling

Formation: Squad in Column Fire Teams in Wedge

Bravo TM Leader

Approx 20 meters

Squad Leader

Alpha TM Leader

Direction of Travel

Traveling Overwatch (9 man squad example)

Traveling overwatch is *used when contact is possible.* This is an extended form of traveling in which the lead element moves continuously but trailing elements move at varying speeds, sometimes pausing to overwatch movement of the lead element. Crew served or special weapons move near and under the control of the squad leader so they can employ quickly. Rifle squads normally move in column or wedge formation. Ideally, the lead team moves at least 50 meters in front of the rest of the element. The trail team and the SL are positioned where they will not become decisively engaged if the unit is attacked or ambushed. This distance allows the SL to direct Bravo Team (B TM) to the enemy's flank should A TM become engaged.

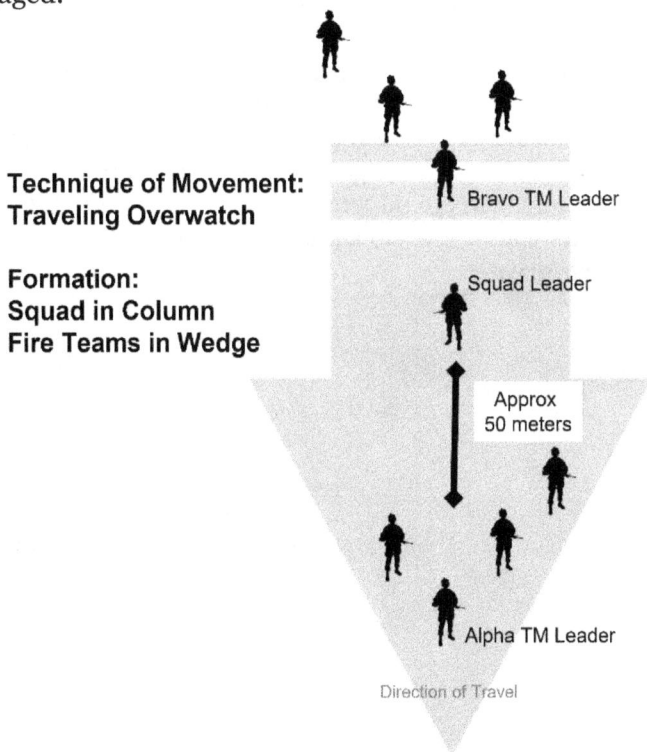

Technique of Movement:
Traveling Overwatch

Formation:
Squad in Column
Fire Teams in Wedge

Bravo TM Leader

Squad Leader

Approx
50 meters

Alpha TM Leader

Direction of Travel

150

The trail element maintains dispersion based on its ability to provide immediate suppressive fires in support of the lead element. The intent is to maintain in-depth, provide flexibility, and sustain movement in case the lead element is engaged. The trailing elements cue their movement to the terrain, overwatching from a position where they can support the lead element if needed. Trailing elements overwatch from positions and at distances that do not prevent them from firing or moving to support the lead element. The idea is to put enough distance between the lead units and trail units so that if the lead element comes into contact, the trail units will be out of contact but have the ability to maneuver on the enemy.

Traveling overwatch requires the leader to control his subordinate's spacing to ensure mutual support. This involves a constant process of concentrating (close it up) and dispersion (spread it out). The desire is mutual support, with its two critical variables being weapon ranges and terrain. Our typical weapon range limitations dictate (generally) we should not get separated by more than 2-300 meters (terrain dependent). In compartmentalized terrain this distance is closer, but in open terrain this distance is greater.

Bounding Overwatch

Bounding overwatch is used when contact is expected, the leader assesses the enemy is near based upon movement, noise, reflection, trash, fresh tracks, unmanned sensors, or experience and transitions from traveling or traveling overwatch to bounding overwatch.

Bounding Overwatch

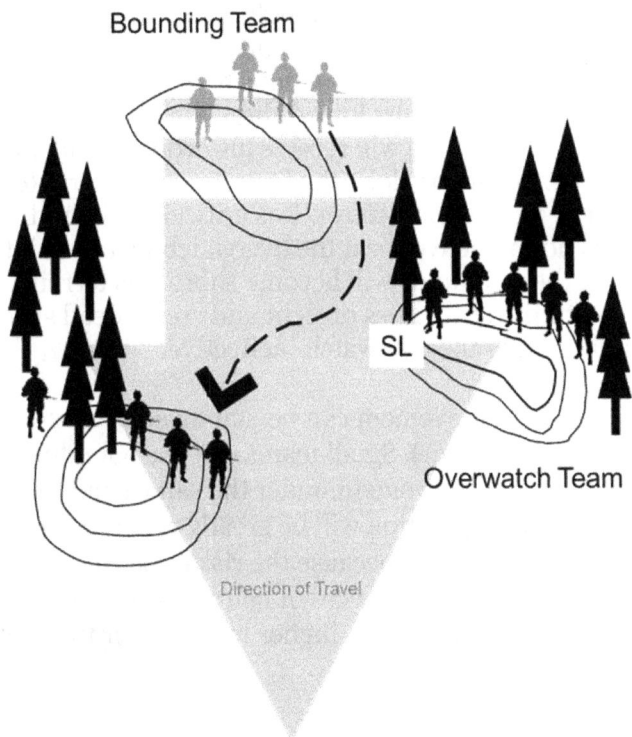

Bounding Team

SL

Overwatch Team

Direction of Travel

He may also use this technique of movement to cross a large open danger area if it cannot be bypassed. The lead fire team overwatches first and scans for enemy positions as the bounding element moves. The squad leader usually stays with the overwatch team to best direct both TLs. The trail fire team bounds and signals the squad leader when his team completes its bound and is prepared to overwatch the movement of the other team ("leap frog"). The bounding element can use traveling overwatch, bounding overwatch, or individual movement techniques (low and high crawl, and three- to five-second rushes by the fire team or buddy teams).

Both fire team leaders must know which team the squad leader is moving with. The overwatching team leader must know the route and destination of the bounding team as well so he can position his overwatch. The bounding team leader must know his team's destination and route, possible enemy locations, and actions to take when he arrives there. He also must know where the overwatch team will be and how he will receive his orders. Available cover and concealment along the bounding team's route dictates how its members move. It is critical the bounding team does not maneuver past the overwatch team's ability to support them. Bounds will become shorter as terrain and vegetation restrict lines of sight and create dead space that would render the overwatch ineffective.

This technique of movement can be employed by all size elements (up to a point). Small teams can use bounding overwatch to cover movement under the same conditions, the tolerance for separation will be greatly reduced for smaller elements. This is because the risk of becoming separated and the consequences of being engaged in relative isolation so are much higher for single or two man elements in a bound.

Individual Movement Techniques (IMT) Under Fire

High Crawl, Low Crawl, 3-5 Second Rush

You will use Individual Movement Techniques (IMT) when moving under direct fire. This ancient graphic on the left is as applicable today as the day it was made; poor choices will have tragic consequences. The sole purpose of IMT is to safely maneuver to a position of advantage against an enemy and engage him. Pick a route that minimizes your exposure to enemy fire and ensure the route does not cross in front of other team members (doesn't mask their fires). Communicate your movement intent to your buddy or team leader using verbal or hand and arm signals. Select the route to the firing positions / cover, a gully, ravine, ditch, or wall at a slight angle to your direction of travel may provide cover and concealment when using the low or high crawl movement techniques. Hedge rows or thick vegetation may only provide concealment (not cover) when conducting IMT. Suppression of the enemy may be accomplished by another element, a buddy, or by yourself. Do not expose yourself to fire unless the enemy is suppressed. With the enemy suppressed you can select an individual movement route or initiate movement. You will move from cover to cover using one of the three proven techniques described on the next page.

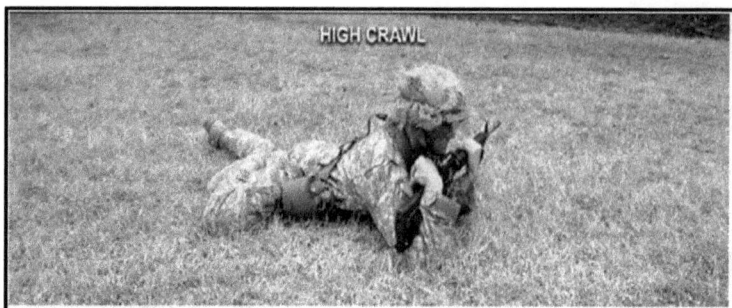

HIGH CRAWL

Using the High Crawl lets you move faster than the low crawl (discussed below) but still gives you a low silhouette. Use this crawl when there is good cover and concealment but enemy fire prevents you from standing up / conducting the 3-5 second rush.

To High Crawl:

Keep your body off of the ground by resting your weight on your forearms and lower legs.

Cradle your weapon in your arms while keeping the muzzle of your rifle off the ground.

Keep your knees well behind your buttocks so you stay low and move forward by alternately advancing your right elbow and left knee, and left elbow and right knee.

LOW CRAWL

The low crawl gives you the lowest silhouette but is the slowest of the three IMTs. Use the low crawl to cross places where the cover and/or concealment are very low and enemy fire or observation prevents you from getting up.

To Low Crawl:

Keep your body as flat as possible to the ground and hold the sling at the upper sling mounting point with your right hand. (your height over ground will be determined by your load carriage choice; mags and kit in front of the hips or on the chest will affect the low crawl)

Let the hand guard rest on your forearm while keeping the muzzle of the weapon off the ground.

Keep your left side of your head or helmet to the ground as you move forward by reaching both arms forward while pulling your right leg forward. Low crawling is pretty awful as you are literally dragging your body across the ground; however it does provide the lowest possible silhouette as you move.

Pull with both arms while pushing with your right leg. Visibility is limited since you are not raising your head up, rotating your head and eyes to check the direction of travel is the only way to see what is to your front. Repeat until you reach your next position.

156

Use the buddy team rush to move quickly from one covered position to another when enemy fire allows brief exposure.

To Rush:

Select your next covered position. Let your buddy know you are getting ready to move so he can be ready to adjust his rate of fire (and ensure you both aren't moving at the same time).

From prone, kneeling, or standing break cover in an unpredictable way. For example, do not rush straight forward from your last firing position or exit from the same side you were firing from as this will allow the enemy to engage you with the least amount of movement of his weapon.

Sprint for 3-5 seconds to your next covered position ("I'm up, he sees me, Im down"). Fast and short exposure is always better for obvious reasons.

To stop and get down quickly plant both feet just before hitting the ground, fall forward and drop to your knees while sliding your firing hand down to the heel of the butt of your weapon and break your fall with the butt of your weapon.

Actions on Contact

Contact occurs when a member of the unit encounters a situation requiring a lethal or nonlethal response to the threat. These situations may entail one or more forms of contact. Contact does not necessarily mean you are being engaged with fires (either direct or indirect). Direct fire is generally the accepted meaning when maneuvering as a small unit, a squad member will not yell "CONTACT!" or report that he is in contact if he encounters an obstacle or civilians. Everything in context always. There are 8 forms of contact by current doctrine, we will use the five below. Electronic warfare etc will still be reported and addressed, but common sense is applied to handle these as discrete enemy actions and not "contact".

- Direct Fire.
- Indirect Fire (receiving mortar or artillery).
- Aerial (aircraft or unmanned systems).
- Obstacles or IEDs
- Visual (can see the threat either through sensors or eyeballs).

Execute actions on contact using a logical, well organized process of decision making and action entailing these five steps:

- Deploy and report.
- Evaluate and develop the situation.
- Choose a course of action.
- Execute the selected course of action (or recommend a course of action to the higher commander).

This five-step process is not intended to generate a rigid, lockstep response to the enemy. Rather, the goal is to provide an orderly framework enabling the unit to survive the initial contact and apply sound decision making and timely actions to complete the operation. Ideally, the unit sees the enemy (visual contact) before being seen by the enemy; it then can initiate direct contact on its own terms by executing the designated COA. The unit treats obstacles like enemy contact, assuming the obstacles are covered by fire. The unit's security force gains tactical advantage over an enemy by using tempo and initiative to conduct these actions, allowing it to gain and maintain contact without becoming decisively engaged. How quickly the unit develops the situation is directly related to its security, and the tempo is directly related to the unit's use of well-rehearsed SOPs and drills.

Leaders understand properly executed actions on contact require time at the unit level. To develop the situation, a squad or platoon may have to execute flanking movements, conduct reconnaissance by fire, or call for and adjust indirect fires. Leaders must give their subordinates time to develop the situation. If you are working duties in command post do not nag the unit in contact for reports; give them time and support as needed. Each of these activities requires time, and the leader balances the time required for subordinate elements to conduct actions on contact with the need for additional combat power or enablers to maintain momentum.

Deploy and Report

If the leader expects contact based upon reports, through reconnaissance, or other means, the formation is deployed by transitioning to the bounding overwatch movement technique. If you are alert to the likely presence of the enemy, your team has a better chance of establishing the first visual and physical contact on its own terms. This contact usually is made by an overwatching or bounding team, which initiates the actions on contact. In a worst-case scenario, a previously undetected (but expected) enemy element may engage the unit. The unit in contact then conducts a battle drill for its own survival and initiates actions on contact. When making unexpected contact, the squad immediately sends a contact report to the next higher element.

Develop the Situation
While the unit deploys, the leader evaluates and continues to develop the situation. The leader quickly gathers as much information as possible, either visually or, more often, through reports of the team in contact and analyzes the information to determine the size of enemy element, location, obstacles, and how he is going to maneuver his team or squad to win. This is all done immediately at the team and squad level. A small unit leader will assess the situation and start directing his element.

Choose a Course of Action
After developing the situation and determining he has enough information to make a decision, the leader selects a COA meeting the requirements of the commander's intent that is within the unit's capabilities.

Execute
In executing a COA, the unit transitions to maneuver. The unit can employ a number of tactical tasks as COA, which may be preceded or followed by additional maneuver. Some of the tasks the squad or team may conduct are attack by fire, bypass support by fire, or disengage. As execution continues, more information becomes available to the leader. Based upon the emerging details of the enemy situation, the leader may have to alter his COA during execution. For example, as the squad maneuvers to destroy what appears to be a dismounted two man enemy team, it discovers two additional squads in prepared positions. The leader analyzes and develops the new situation. He then selects an alternate COA, such as establishing a support-by-fire position to support another platoon's maneuver against the newly discovered enemy force.

React to Ambush

React to ambush is required when the team or squad enters a kill zone and the enemy initiates an ambush with a casualty-producing device and a high volume of fire, the unit takes the following actions:

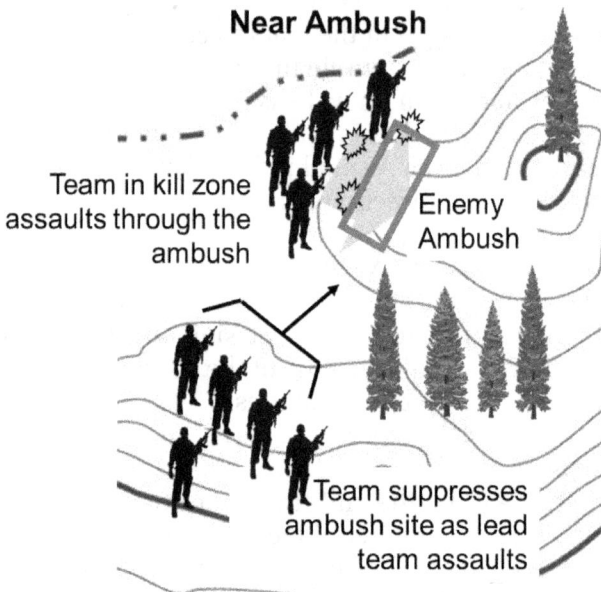

Near Ambush

Team in kill zone assaults through the ambush

Enemy Ambush

Team suppresses ambush site as lead team assaults

Near Ambush. In a near ambush (defined as being within hand-grenade range, about 30 yards), individuals in the kill zone immediately return fire and assault through the ambush using fire and movement without orders. (In the unlikely case that fragmentation grenades are available take up covered positions, and throw. Immediately after the grenades detonate, assault through the ambush). Team members not in the kill zone locate and place well aimed suppressive fire on the enemy. They take up covered positions and shift fire as the assault begins. Use smoke or flash bangs to assist with confusing or disrupting the ambushing enemy.

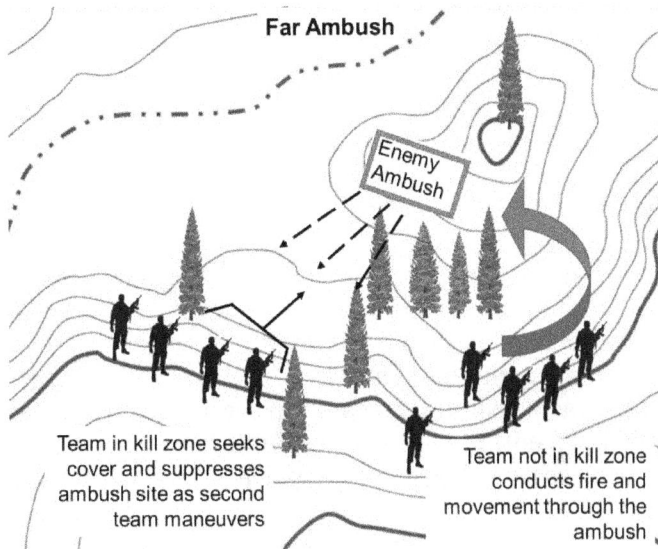

Far Ambush

Enemy Ambush

Team in kill zone seeks cover and suppresses ambush site as second team maneuvers

Team not in kill zone conducts fire and movement through the ambush

Far Ambush. In a far ambush (defined as being beyond hand-grenade range, about 30 yards). Team in the kill zone immediately returns fire and takes up covered positions. A fire team may be split and squads may become comingled, just roll with it and fight through the problem. The SL identifies the enemy's location and shooters place accurate suppressive fire on the enemy's position. Attempt to target and engage enemy crew served weapons first and gain fire superiority as quickly as possible. Friendlies not in the kill zone begin fire and movement to destroy the enemy. The assaulting team/squad fights through the enemy position or you force the enemy to withdraw. The other option is your unit can disengage and reorient to continue your original mission. This is all dependent on commander's intent and the enemy situation. The leader reports to higher, reorganizes as necessary, and continues the mission.

React to Indirect Fire (IDF)

While the unit is moving or stopped in unprotected
positions, indications of incoming rounds occurs. Any team
member announces, "INCOMING!"; team members
immediately assume prone or nearby covered positions
during the initial impacts. After the initial incoming rounds
impact, your leader determines the extent of the impact
area and the nearest edge out of it. Then, rapidly applying
METT-TC he gives the direction and distance to move out
of the impact area to a rally point (still heading roughly in
the direction of travel if possible; the direction of travel is
always 12 o'clock no matter what cardinal direction you are
moving) for example he may yell, "Two o'clock, three
hundred meters". You and your team members will repeat
and relay the command and quickly move (run). Stay as
organized as possible and maintain security/scanning as
best you can enroute to the rally point; at a minimum
buddy teams must keep track of one another. Once at the
location the team leaders will get ACE reports from all
members, reorient, and then continue mission.

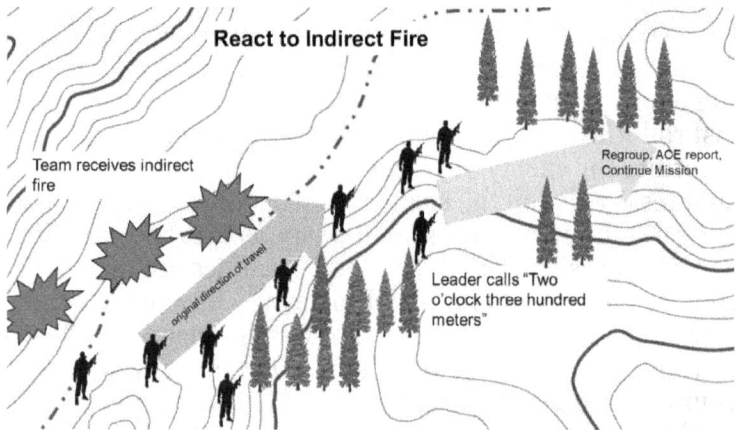

React to Indirect Fire

Team receives indirect fire

Regroup, ACE report, Continue Mission

original direction of travel

Leader calls "Two o'clock three hundred meters"

React to Unmanned Aerial System (UAS)

Any contact with unidentified UAS should be reported immediately. If you are in the field and encounter an unknown UAS of any size or action it must be assumed that the intentions are hostile; the risk to the unit is too great to assume otherwise. Visually spotting or hearing a small UAS indicates the enemy control station is and associated enemy forces are close by. Give the air attack alarm by hand and arm signal, verbal, and over tactical nets. Occupy defensive positions with cover and concealment if available. When making the decision of whether to fire at non-attacking hostile aerial platforms with small arms,

LSS Drones are a significant threat to irregular forces

take into consideration the assigned mission and tactical situation and know that UAS are *very* difficult to defeat using direct fire weapons. Once the decision to engage has been made, continue to scan for additional enemy UASs as the engagement takes place. Your SL or patrol leader will make the engagement decision and your unit engages the aerial platforms with all available rifles and machine guns (if available). Expect the firing signature from small arms to disclose the unit's own position, so METT-TC must be ever present in the decision process. Once the lead distance is estimated, the riflemen and machine gunners (if available) aim and fire their weapons at the aim point (a distant feature that everyone can aim at) in front of the

UAS until the aircraft has flown past that point. Maintain the aim point, not the lead distance as you are "ambushing" the flight path and shooting in one spot. Every weapon must be used to engage the target with the goal of placing as many bullets as possible in the enemy's flight path. The weapon should not move once the firing cycle starts. This type of engagement is the best of the terrible options available to a rifle squad for engaging a small UAS. Adding shotguns into formations may be suggested, however the range and payloads are inadequate and would not be worth the added weight to carry one. It is a non-solution. Another last ditch option is to attempt to destroy or degrade the UAS sensor package by directing high power lasers directly at the sensors (again we are discussing LSS type, "Low, Slow, Small" UAS). This entire drill is related to these types only; high altitude exotic sensor suite UAS are a different animal altogether. Passive measures of camouflage, route planning, leveraging weather advantages, etc are all better than having to engage a UAS that has spotted you and is now actively hunting your unit.

React to UAS

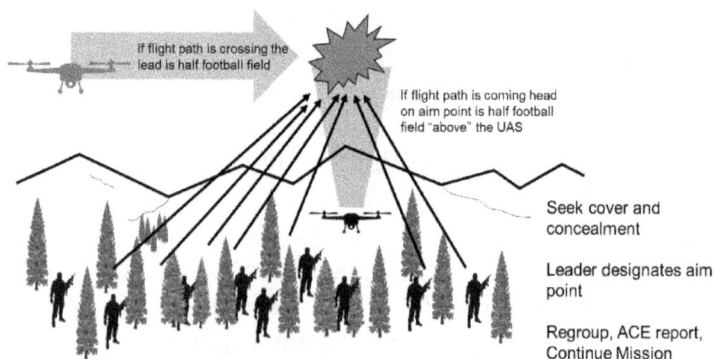

If flight path is crossing the lead is half football field

If flight path is coming head on aim point is half football field "above" the UAS

Seek cover and concealment

Leader designates aim point

Regroup, ACE report, Continue Mission

Chapter 4
Communicate

Our community and society at large has become desensitized and somewhat spoiled with regard to the ability to instantaneously communicate on demand. We can access detailed data on obscure subjects, message loved ones with annoying regularity, or call a friend who lives halfway around the world. Emergency services are only three digits away; we can reach out for assistance at any time with relative confidence that we will receive it. Communications are so common and so much a part of our lives we have come to take them for granted. Disruption of comms in any way, even if limited and local, can add to or even cause chaos very quickly. The loss of modern comms will be one of the most difficult adjustments we will have to make, but there are alternatives that can be practiced beforehand.

This section is an overview of concepts and highlights of using radio and visual comms. Radio communications are a complex task and it can take years to become an expert. Most of us want to just pick the radio up, put it on "Channel 6" and talk to anyone we want to. Unfortunately, this is not how it works and it takes a lot of prep work and practice to use the technology in a manner that meets our requirements.

Visual Signals

Visual signals are any means of communication that require sight and can be used to transmit prearranged messages rapidly over short distances. This includes the devices and means used for the recognition and identification of friendly forces. The most common types of visual signals are hand-and-arm, flag, pyrotechnic, and ground-to-air signals. However, we are not limited to the types of signals discussed and may use what is available.

Chem light sticks, flashlights, and other items can be used provided their use is standardized within a team or region and understood by Citizens working in the area. There are multiple dozens of hand and arm signals, far too many to address in detail. Some of the more common ones you will need to know are below, your team or group will be the deciding factor for which ones are in common use for your area. When selecting non-standard hand and arm signals to complement or supplement we recommend not duplicating and changing the meaning of standard (Army or USMC) signals. This can create confusion within teams.

If it is not feasible to perform the proper hand-and-arm signal while holding your firearm, an alternate signal should be used with the non-firing hand. In the event that you must use both hands to perform a signal, the weapon will be moved to the hang or sling arms position.

Wedge extend the arms downward and to the sides at a 45-degree angle below the horizontal, palms to the front. Alternately, use the non-firing hand and point the index finger and pinky finger to the ground.

What you do – execute a change of formation drill IAW your team SOP. Pass the signal on to those behind and to your flanks. Flow toward the next position over the terrain and maintain your spot in the wedge.

Assemble or *Rally* raise the arm vertically overhead,
palm to the front, and wave in large, horizontal circles.
Note. When followed by pointing designates an
assembly or rally point.

What you do – assemble / move toward the leader when
this signal is given. If the context is designating an en route
rally point the signal will be followed by knife hand toward
the actual area. For en route RP every member of the unit
will pass the signal on to the teammates behind or to the
flank.

Numbers - All number signals will be given with the non-
firing hand. For numbers greater than nine, the signal will
be given as individual digits in the number. For example,
the number 19 will be represented by the signal for the
number, 1, followed the signal for the number, 9.

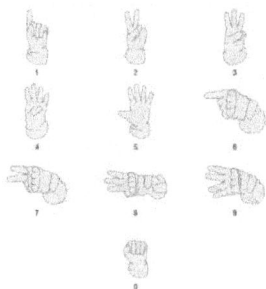

What you do – no requirements other than understanding
the system.

Line Formation - extend the arms parallel to the ground.

What you do – execute a change of formation drill IAW your team SOP. Pass the signal on to those behind and to your flanks. Flow toward the next position over the terrain and maintain your spot in the line.

File (Column) Formation - move the non-firing hand to touch the rim of the headgear directly in front of the face. Fingers will be extended and

What you do – execute a change of formation drill IAW your team SOP. Pass the signal on to those behind and to your flanks. Flow toward the next position over the terrain and maintain your spot in the column as you pick up scanning responsibility for your position.

Contact Left (or Right) - extend the left (right) arm parallel to the ground. Bend the arm until the forearm is perpendicular. Repeat

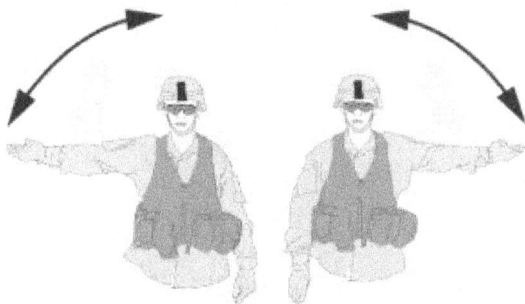

What you do –direct fire contact location may or may not be obvious when it happens. The unit members may use this to indicate direction of the contact if it is not obvious or if they cannot communicate verbally. Execute the associated drill based on the available information and terrain.

Enemy (In sight) - hold the rifle in the ready position at shoulder level. Point the rifle in the direction of the enemy. Alternately, use the non-firing hand, point index finger at the enemy and thumb pointing down. All other fingers will be curled.

What you do – First and foremost do what your team leader or SL does; move, shoot, get down etc when this signal is used. Execute actions IAW the mission and the commander's intent. Seeing the enemy first does not necessarily equate to engaging or ambushing, it is mission dependent.

Freeze - To signal "freeze," raise the fist to head level.

What you do – freeze where you are, make no further movement, and be absolutely silent. If you are mid step place the next footfall quietly to keep your balance and stay there. Do not take a knee, do not adjust your gear. Trust your team mate that called the freeze, he did it for a reason.

Halt - To signal "halt," raise hand to head level, fingers extended and joined.

What you do – stop where you are, face your primary sector of responsibility and pull security. Make no further forward movement other than what is necessary. Depending on unit SOP you may take a knee immediately, after a set period of time, or on command from the TL/SL. Halts may be for a few seconds or for extended periods of time, so stay alert for changes or guidance from your leadership...and stay awake. Do not adjust equipment, drop rucks, etc.

Danger Area - To signal "danger area," draw the right hand, palm down, across the neck from left to right.

What you do – this will get a bit more complex since this hand and arm signal starts a sequence of actions for the entire unit. You will have a specific job or tasks for crossing a danger area. You will execute based on SOP and what position you have in the squad or team.

Pace Count - To call for a pace count, tap the heel of boot repeatedly with an open hand.

What you do – this is when the leader calls for the pace man to report current pace count in distance (meters). You pass the signal back in the formation, or you report the count to the patrol leader if you are the pace man.

Head Count - To call for a headcount, tap the back of the head repeatedly with an open hand.

What you do – this is when the leader calls for the head count. You pass the signal back in the formation and wait for the count to come back forward (typically used in a column / Ranger file). This can be the most frustrating thing ever for a leader at night with a bunch of tired dudes. Sounds like an easy task, but getting an accurate head count has been known to make grown men cry in the woods. When the count comes back up (starting from the rear) the man behind you will grab your arm or shoulder and quietly say a number. That is all he will say. If he says "eleven" that means you are number 12. That is what you pass up to the next man, just *your* number. You say "twelve". You don't tell him "hey I'm twelve and you are thirteen". Just say your number, stay awake, and the world will turn. You may not be the same number during various headcounts on the same patrol due to formation changes etc.

Categories of Communications

Normal State Communications (NSC). This category includes the typical consumer communications we use on a daily basis. Mobile phones, internet access, social media. And even old school land line phones all fall into this broad category. These are open source, easily disrupted, easily monitored, and readily collected on. Weather events, power grid disruption, or intentional interference / denial of service can and will render these unusable.

Categories of communications in a non-mobile phone world:

Contingency Communications (sometimes referred to as Sustainment Communications in the community). Comms in this category are intended as an alternative to the NSC options. Examples of these include the use of GMRS or ham radios in the absence of mobile phone or internet service. Contingency communications are meant to fill the gap left when the NSC options have been disrupted by an event or are simply unavailable due to remote locations without internet or mobile service. Ham radio is a great example of contingency communications use. While the ham community leaves a lot to be desired many times (and can be *very* difficult to tolerate) they are the prime example of a group that practices and uses contingency communications on a daily basis. A proficient ham radio expert can be an excellent asset to Pro Citizen groups, building proficiency with the ham radio systems is something you can do well before the skillset is needed. It is one of those skills that cannot be self-taught in a crisis, we highly recommend that radio comms is a skill you prioritize.

Tactical Communications. These are military type communications in which information of any kind, especially orders and reports conveying actionable information are conveyed between units, individuals, or facilities, particularly during combat operations over short(er) distances. It includes any kind of delivery of information, whether verbal, written, visual or auditory, and can be sent in a variety of ways. Tactical communications are time sensitive and immediate as they will contain information or orders / directives that require action. Prevention of interception or detection are paramount. Using high loss "stubby" antennas, low power, directional antennas coupled with terrain masking, narrow transmission windows, and coded, short transmissions are all necessary to prevent compromise.

Strategic Communications. Strategic comms are high level communications that direct and pass information of significant importance. The intercept or loss of these type of communications will result in substantial damage to the organization, as such these must be encoded / encrypted and covert in nature. These messages and data are instrumental to directing and coordinating the actions of the greater organization and are typically larger in scope and longer in distance than Tactical Communications. HF, encryption, deception, and burst data are all practices to protect these transmissions. This manual will not address Strategic Comms or the methods to encode transmissions; there are other outstanding detailed references available in the community that describe these procedures.

Hardware

Ham radios. Also known as amateur radios, these are versatile and portable communication devices that can be used in a variety of settings, including Contingency and Tactical Communications. There are base station, mobile (vehicle mounted) and handheld ("handy talky" or HT). These radios can be built to receive and transmit on HF, VHF, and UHF bands, ensure the radio you choose fits with your and your group's requirements. The capability to communicate without mobile phones or internet comes at a price though. A transmission will have two components; the carrier signal (the electronic signature) and the message itself (information). The signal can be detected, used for direction finding, or interfered with (jammed). The message itself can be intercepted, used to build intelligence against friendly forces, or can be mimicked to issue false orders. The sophistication of AI capabilities is growing at an exponential pace; duplicating and deep faking a friendly voice on the radio to cause confusion or disrupt operations will be an easy task. Citizens must be disciplined in following radio procedures to ensure our messages are not being intercepted. We must take actions to minimize the chances of our transmission signals being detected. Use of encryption, coded messages, and timing / net discipline are all viable countermeasures against these threats. No system is foolproof, understanding the vulnerabilities and having SOPs in place before X Hour will help the team mitigate the risk of using radio communications.

Choose the Right Radio. The ideal radio for the Pro Citizen is an HT that is small, lightweight, and durable. It should have a low power output option you can use to reduce the risk of your transmission being detected. Some of the most popular handheld ham radios for tactical situations are the Yaesu VX-6R, Baofeng UV-5R, Baofeng AR152, and the Icom IC-T70A. These are not necessarily recommendations; they are just examples of what we currently see in the community. The addition of a dedicated handheld receiver for scanning and monitoring common HF and shortwave as well as FRS/GMRS/MURS traffic is also an option that will allow dedicated monitoring without using your primary HT. These should be considered for your dedicated RTOs (Radio Telephone Operators) in your teams. Radios need to be protected or modified to prevent switches and screens from being turned on or off at inopportune times. A radio pouch or component of your chest rig can easily key an exposed PTT (Push to Talk) switch; having a hot mic on the squad or platoon net is preventable pain. It is almost a rule that the net will get hot mic'd at the worst possible time; identifying and correcting radio issues during PCCs and training is critical for the team.

Use the Right Antenna. The antenna you use with your ham radios will greatly impact its ability to transmit and receive signals and can alleviate *some* risk of intercept. In a tactical situation, it is important to use a compact antenna that can be easily concealed. Some of the most popular compact antennas for handheld ham radios are the Nagoya NA-771 and the SMA-Female Dual Band Antenna. The short, stubby antennas for dual band, analog radios (Baofeng UV5R type) are perfect for tactical communications. bringing multiple antennas, connections, and of course batteries, when going into the field to maximize your capabilities.

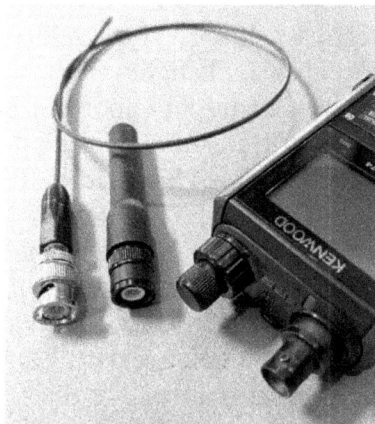

Quick Tip: BNC Adapters all around will let you attach various antennas quickly. Antenna interfaces can be fragile, a BNC can be a decent preventative measure.

Do not use memory channels. A preprogrammed HT is a treasure trove of info for the enemy if it is captured. Frequencies should be punched in for each mission and should change daily or by mission requirements IAW the frequency plan of your group. Having a memory programmed HT for daily use and practice, weather events, community support, or local emergencies is good to go and is recommended. Just be sure to train your team to operate without programming your comms; we must be disciplined enough to follow group SOPs and keep vulnerable data to a minimum. Once X Hour occurs, we must have empty memory channels in our comms systems to maintain COMSEC and have a plan to Z Out (delete all data) or destroy radios if they must be abandoned or are at risk of capture.

General Rules for Radio Use

Use established formats to expedite transmissions such as sending reports (SALUTE, SALT, SITREP etc).

Use the phonetic alphabet and numbers.

Transmit clear, complete, and concise messages. When possible, write them out beforehand.

Always assume you are being monitored and DF'd

Speak clearly, slowly, and in natural phrases as you enunciate each word. If a receiving station must write the message, allow time for him to do so.

Listen before transmitting to avoid interfering (stepping on) other transmissions.

Use prowords to reduce transmission time and avoid confusion.

Minimize transmission time; key the mic, send your traffic, and get off. PTT is "Push to Talk"; the running joke in training is for some it is a "Push To Think" button. *Know what you are going to say before you key up.*

Do not violate radio silence when RLS (Radio Listening Silence) is in place.

Do not make unnecessary or long duration transmissions.

Do not identify an individual by name or any other personal information.

Do not speak faster than a station experiencing poor reception conditions can be expected to receive.

Control your emotions. Panic is contagious, a calm voice on the radio will always be of benefit to the team.

Use Short and Infrequent Transmissions. When using a handheld ham radio in a tactical situation, it is important to keep your transmissions short and infrequent. This will reduce the risk of your transmission being detected and minimize the amount of time you are on the air.

Use Coded Messages. To further reduce the risk of your transmission being detected, it is a good idea to use coded messages (not legal under current FCC rules, not advocating doing so under rule of law conditions). This can be done by using simple substitution ciphers or by using a more sophisticated encryption system such as One Time Pads (OTP). When using coded messages, it is important to ensure that all parties involved in the communication understand the code.

Use the Right Frequency. When using a handheld ham radio in a tactical situation, it is important to use the right frequency. In most cases the typical handheld (HT) radios use a low-frequency band, such as the 2-meter band or 70-

centimeter band. Odds are these frequencies in these bands will be highly monitored and targeted by any threat. Some HTs have the capability to operate in the 220 mhz range (the 1.25 meter band); this band is still vulnerable to intercept and jamming however the relative obscurity may provide an opportunity to mitigate these threats. Unsophisticated near peer threats and competing threat groups will potentially not focus their collection or monitoring efforts on this band (as much). This does not relieve you from using correct radio procedures and countermeasures but it may provide a better option than the standard 2 meter and 70 cm bands.

Avoid Broadcasting Your Location. When using a handheld ham radio in a tactical situation, it is important to avoid broadcasting your location. This can be done by using directional antennas and avoiding giving out your location in your transmissions. Never transmit a message that contains unencoded friendly locations. The second component of this concept is DF (direction finding) avoidance or mitigation. Tactical communications within and between nearby units should be infrequent and use the lowest possible power with the most "lossy" antenna that still works (term for an antenna that is inefficient and therefore has less of a chance to carry a signal outside of your area). Comms windows for transmissions back to command posts or leadership organizations must be planned ahead of time to further minimize the exposure time and frequency of transmissions. Comms locations must also be taken into consideration for these types of longer range transmissions. Use terrain masking for planned communications. The frequencies we will use are typically line of sight, so choosing a transmit location that has a higher terrain feature between you and the threat is an effective way to cut down the chances of being discovered. Do not send these types of reports or transmissions from a patrol base, LP/OP etc back to a distant unit or command post, always move to a location at least 200 meters away from static friendly forces.

Monitor the Airwaves. When using a handheld ham radio in a tactical situation, it is important to monitor the airwaves for any suspicious activity and adversary traffic. This can be done by listening to the background noise on the frequency you are using, as well as listening for any unusual signals.

Typical handheld (HT) radios for the Pro Citizen. Kenwood D74 (discontinued but good used units can still be found), UV5R with an extended battery, and GT3WP (essentially a UV5R with a bit more durability and some water resistance).

Phonetic Alphabet (NATO Standard)

Numerals will be transmitted digit by digit except multiples of thousands may be spoken as such. For example, 17 would then be One Seven (pronounced WUN – SEV-EN).

A - ALPHA	N - NOVEMBER	1 - WUN
B - BRAVO	O - OSCAR	2 - TOO
C - CHARLIE	P - PAPA	3 - TREE
D - DELTA	Q - QUEBEC	4 - FOW-ER
E - ECHO	R - ROMEO	5 - FIFE
F - FOXTROT	S - SIERRA	6 - SIX
G - GOLF	T - TANGO	7 - SEV-EN
H – HOTEL	U - UNIFORM	8 - AIT
I - INDIA	V - VICTOR	9 - NIN-ER
J - JULIET	W – WHISKEY	0 – ZERO
K - KILO	X - X-RAY	
L - LIMA	Y - YANKEE	
M - MIKE	Z – ZULU	

Procedural (Pro) Words

Acknowledge: Use this when the person you are addressing must acknowledge receipt of the message.

Affirmative: Yes, or that is correct.

Break break: You have an urgent message and need to interrupt the current conversation

Correction: Indicates that an error has been made and that the transmission will repeat from the last word correctly used.

I say again: I will re-transmit the message, or part of the message

I spell: The word will be spelled using the phonetic alphabet

Negative: No, or that is not correct

Out: End of transmission – no reply is expected

Over: means it is the end of transmission and a reply is expected.

Prepare to Copy: I will send the message when you are ready, usually a "heads up" that a report or detailed message will follow.

Radio check: what is my signal strength and readability? "You (callsign) this is me (callsign), radio check, over"? If clear signal receiving station answers with "you this is me callsign" and "Roger Over". You reply with "roger out". That is all you say unless the signal is poor, then you describe the issue.

Relay to: Transmit this message to the third party callsign as indicated "Relay to BLUE 27"

Roger: Message received and understood

Say again: Please repeat your last transmission

This is: Indicates the calling unit's identification is next. For instance, if C6 were making a call and needed to identify themselves, they would say "This is CHARLIE SIX"

Wait: A pause of a few seconds follows

Wilco: I will comply with your message

There are several categories of communication types and/or purposes that have been created in the community; for our use we will group them into just a few for simplicity's sake.

Standard Report Formats

Always keep in mind when you are building and sending reports it serves a purpose. The "so what?" of reports we send, receive, or call for must be kept in focus. Units get in a bad habit of calling for reports just because. Always have a purpose when on either end of the process. Teams and groups should adopt standard report formatting for typical reports we use. Adopting a standard will ensure uniformity across organizations when groups train or execute missions together. Reports may be written, verbal, transmitted over radio, or taken by runner. The formats remain the same no matter the means of delivery. Some are immediate in nature such as the ACE report, some are routine, and some are event driven such as the SALUTE Report. There is a multitude of report formats, to the point of being ridiculous. Keep it simple, keep it usable in your organizations. Calling for a "Blue 27, Line A4" to be sent can be cumbersome and counterproductive, fortunately we are not restrained to doing so. Get to know the basic reports first and be an expert at sending them over the radio quickly and clearly. Below is an introduction to just a few of the available common format reports you will most likely use.

SPOT REPORT (SALUTE Report Format). Use **SALUTE** to organize and convey concise information about an observation, or a SPOTREP. More info is usually better, remember you are collecting information, not intelligence. Don't try to interpret what you see for a report, just send the facts. Intelligence is a processed, fused, usable product created from the accurate information that is sent to command posts or cells by individual scouts/team members. You will be on the front end of the process, the only way to get a clear picture of what the threat is doing is to have accurate, timely information that can be processed and turned into actionable intelligence.

S- Size Report the number of personnel, vehicles, aircraft, etc. Do not interpret the scene, provide exactly what you see to eliminate the chances of misinterpretation. For example we report "a group of 30 armed individuals" not "a platoon size element"

A-Activity Report detailed account of actions, for example, direction of movement, troops digging in, indirect fire, type of attack, etc. "emplacing mines on roadway"

L-Location. Report where you saw the activity. Include grid coordinates or reference from a known point such as an intersection including the distance and direction from the known point (not your or any other friendly location). *Never send enemy locations in relation to friendly maneuver control measures (graphics) or friendly locations.* The enemy knows where they are, all they need is an intercepted message to put the pieces together and find your location. For example do not report enemy location as "two thousand meters southwest of my location." A quick back azimuth and some seventh grade math at the enemy FDC (Fire Direction Center) and here comes the boom.

U-Unit or Uniform. Report the enemy's unit markings if seen, at a minimum report the uniform pattern (multicam with punisher logo patches) any distinctive features, such as uniforms, patches or colored tabs, headgear, vehicle identification markings, etc.

T-Time. Report the date time group (DTG) time the activity was observed, not the time you report it. If reporting across time zones ensure you specify.

E-Equipment. Report all equipment associated with the activity, such as weapons, vehicles, tools. If unable to identify the equipment, provide as much detail as you can so an identification can be made by higher headquarters. Do not guess! Don't misidentify a piece of equipment, small details matter. Something minor can confirm or deny a particular enemy course of action (COA). A formation of six self-propelled artillery called in as "tanks" will paint an inaccurate picture of the enemy for your leadership.

At the end of each Spot Report ensure you add what your actions are. It may simply be "continuing to observe" but always include what you are doing or plan to do.

SALT – The SALT report (Size, Activity, Location, Time) is reported as a rapid update to your recent SALUTE report. Eg additional threat vehicles or enemy soldiers arrive at a location you are observing. The initial SALUTE report is now inaccurate and should be updated with the additional information.

ACE – **A**mmo, **C**asualties, **E**quipment. Used to report status after contact with the enemy. This is given by SOP or upon demand by a first line leader. For example after an engagement your TL will call for an ACE report to assess status of his team and then report the rolled up status to the SL. Sometimes you may hear "LACE" which adds liquids, this part is usually used in a training environment but may be adopted by your leadership. ACE is reported as **A** the number of full primary mags, **C** any wounds received, and **E** that you have accountability of all sensitive and critical equipment, and they are (readily observed as) serviceable. A status of four full M4 mags, no casualty, all equipment is apparent FMC and present is reported verbally to the TL as "FOUR, UP, UP!" Local and team SOPs may adjust this practice.

Slant (or combat power) Report. This is fairly unit specific and is focused more at the command post level for battle tracking. Units may adjust this for operations, it is generally a way of passing on unit combat power status (in terms of vehicles, systems and combat effective personnel) so a commander can make decisions based on the information. On open nets without electronic encryption this data must be concealed through the use of codes or one time pad (OTP) use.

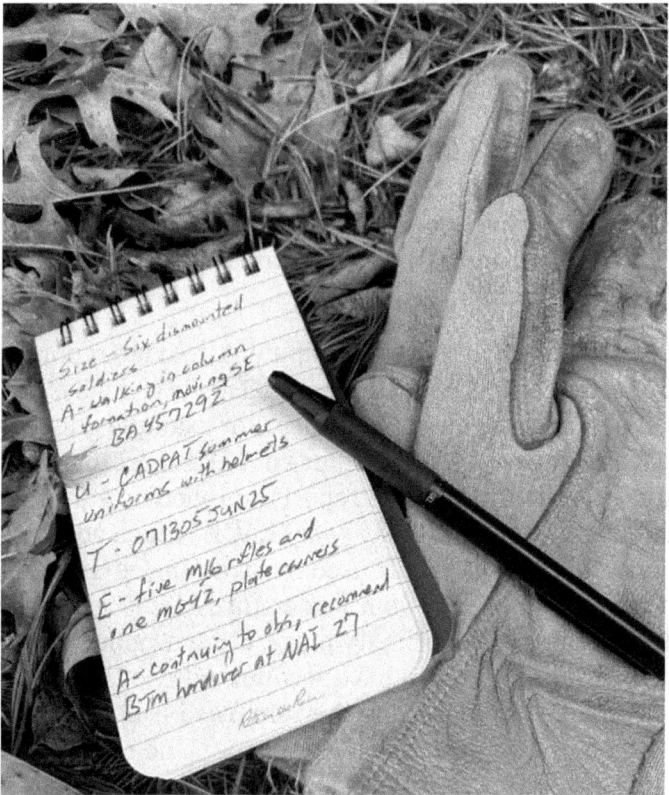

Quick Tip: For reports with any detail write them out before you key up the radio. Have messages encrypted before you push to talk. Be calm, be brief, be clear, and be gone.

Chapter 5
Survive

Survive for our purposes is not just in terms of the skillsets that would support one surviving alone in the wilderness in an emergency. Surviving for a Pro Citizen in the spirit of this manual is remaining combat effective to stay in the fight...and doing so for the duration of a long-term crisis. Keeping your family, team, and yourself safe, sheltered, and fed over the long haul. Medical skills, sound tactical decisions, and basic human needs such as food, water, and shelter are all requisites. These will all be significant challenges in an environment of reduced or non-existent system supported logistics. Having a plan and the competences to fill in the gaps left by a system collapse requires a lot of work before the fact. Being capable of demonstrating your skills during practice and training today will boost your confidence in your ability to call upon those skills when the need arises. The goal of preparation is to build confidence in your abilities and to function despite your situation and fears. Failure to prepare yourself physically and psychologically to cope with the loss of normalcy leads to reactions such as depression, carelessness, inattention, loss of confidence, and poor decision-making. There is an old adage "the more you sweat in peace, the less you bleed in war". Being as prepared as possible before the fact will go a long way to mitigating the negative aspects, both physical and psychological, of what may be coming your way.

Remember that your life and the lives of others who depend on you are at stake. It includes everything from simple wear and care of your outdoor protective clothing and equipment all the way up to application of small unit tactics that will limit risks during direct and indirect fire contact (addressed in Chapter 3). This section will introduce you to the fundamentals and lay the groundwork for further development of some of these skills. As with all things in this effort consult and get professional in person training for these skills. You can fake the funk and get lucky for some things on the tactical side of the equation. But you can't fake food, water, or medical treatment; we must know what we are doing for real when it comes to these.

Rule of "threes". The rule of threes is not a rule per se, it is simply a guideline that the community has to prioritize survival training and equipment choices. They are based loosely (guidelines) on:

Three minutes without air / oxygenated blood. Arterial bleeding can kill in a couple of minutes, and being without air for three minutes can cause irreparable physical damage or death.

Three hours without shelter in killer weather. Low temperatures, rain, snow, ice, wind will heighten the need for adequate shelter. You will only have about three hours to shelter yourself and others in the toughest conditions. Hypothermia, frost bite, heat exhaustion, heat stroke, or death can all result from exposure. Proper clothing or lack thereof will either delay or accelerate the shelter timeline.

Three days without water. Again, this will vary for individuals depending on physical fitness, prior hydration, weather conditions, physical exertion...the factors are many. Some can endure more than three days without water, some will be combat ineffective much faster. Dehydration can progress to heat exhaustion and eventually heat stroke in hot environments. As with many things in our world "it depends" is usually the answer.

Three weeks without food. The situation we are training for will be a calorie consuming beast. The body can survive well past three weeks without food (as long as there is water) but there is an unacceptable limit of performance that will be reached. To be a functioning team member under stress in a physically demanding environment you need calories. This will vary by individual factors such as physical fitness, body mass, fat percentage, and the level of exertion.

<u>*Three minutes*</u> *without air / oxygenated blood*

First Aid and Injury Prevention

*Go **take in person classes!** CPR, First Aid, a high-quality trauma casualty care class...take them. A manual alone will do a disservice to this task. We will address some very surface level concepts in here (and a few tasks), this is a subject that must be trained in person by seeking out experienced professionals. The volumes of information cannot be captured in this manual, this is just a sample / overview of concepts.*

Survival medicine employs four techniques to facilitate unsupported performance of survival proficiencies. Assess these fundamentals during planning and preparation activities and include relevant physical and psychological factors. Additionally, they form the basis for effective organization, training, and equipping efforts by the community or MAG leadership, unit recovery force and the individual.

Prevention. The common sense act of proactive prevention of an action that could lead to a requirement to perform trauma first aid or survival medicine. For example, during movement, look for alternate crossing points across a river instead of swimming across to decrease the chance of getting hypothermia and having to dry your clothing.

Recognition. The act of actively recognizing:

Symptoms (urine color and relation to

dehydration).

Capabilities (identify local medicinal plants such as cattails, willow in your area of operations).

Identity (poisonous snakes or insects or hazardous plants).

Mitigation. The actions taken to immediately reduce the severity and or pain associated with a survival related injury or illness (application of a tourniquet to stop severe, uncontrolled bleeding that could cause loss of life)

Treatment. The actions taken to manage and care for an unsupported person with a disease or disorder and restore their health. The ability to plan, prepare, execute and assess appropriate actions to meet your survival needs significantly increases their overall chances of survival, increases morale, and aids in individual and the team's ability to perform the survival proficiencies and remain combat effective.

Personal Hygiene and Sanitation

The community and team can avoid many kinds of illnesses and infections by practicing good sanitation and hygiene. Maintaining a clean body and living area will prevent the spread of germs and bacteria whether alone or in a group. It will also allow individuals and patrols to stay organized and protect items from animal contact.

Application of the following guidelines regarding personal health and hygiene will enable unsupported individuals and teams to safeguard personal health and the health of others while isolated from normal services (water, sewer, electricity etc).

Stay clean (daily regimen) and minimize infection by washing. (Use white ashes, sand, or loamy soil as soap substitutes.) Clean debris from your hair; keeping hair cut short will assist with hygiene, but dependent on METT-TC and blending in may require adjustments to grooming. Wash your hands thoroughly after handling any material that is likely to carry germs, after urinating or defecating, after caring for the sick, and before handling any food, food utensils, or drinking water.

Oral hygiene will prevent long term issues that could literally take a troop out of the fight later on. Inspect hygiene kits routinely and ensure you have the needed items to keep up with this.

Clean and protect your feet. This cannot be overstated. A dismounted irregular force is 100 percent reliant on their mobility, that starts with a good pair of boots with durable healthy feet inside them. Change and wash socks routinely, acquire several good sets of outdoor / hiking socks and rotate them out as they wear. Wash, dry, and massage daily and check frequently for blisters and red areas. Train now by conducting long movements over terrain and get some ruck miles in every week. Use adhesive tape/mole skin to prevent damage.

Keep your clothing and bedding as clean as possible to reduce the chances of skin infection or parasitic infestation. Prevent and control parasites and check regularly for lice, fleas, ticks. Pick off insects and eggs (do not crush).

Combat Casualty Care Overview

Survival medicine is characterized by remote and improvised care in an environment with little to no external resources, with routine or exotic physical or psychological illnesses or trauma, and delayed or nonexistent evacuation to definitive (hospital) care. Because medical personnel will not always be readily available, nonmedical team members must rely on themselves and other teammates' skills and knowledge of first aid methods to render aid. This chapter provides first aid procedures for nonmedical personnel in environments from a peaceful community affected by a natural disaster to combat situations. Your team or MAG should integrate individual and multiple first aid tasks in combination with collective tasks into various training scenarios.

Excluding the enemy, medical related problems arising from combat and isolation pose the greatest threat to you and the team. Your team must be able to perform fundamental survival medicine techniques throughout the duration of combat operations. Follow guidelines and protocols focused on the care of casualties in a combat or tactical environment at the point of injury. You must be prepared to provide self-aid or buddy-aid in the absence of a medical provider.

Survival medicine recognizes the fundamentals of the MARCH acronym below and provides a flexible approach that prioritizes actions relevant to physical, psychological, and environmental considerations that are continually assessed and prioritized for performance by fighters with limited support.

Follow the **MARCH** acronym as a guide to the sequence of treatment priorities in caring for any combat casualties:

Massive hemorrhage — control life threatening bleeding. The number one potentially survivable cause of death at the POI is hemorrhage from a compressible wound or any life-threatening extremity bleed. The rapid application of a quality tourniquet (CAT and SOF-T are two proven designs) is the recommended management for all life-threatening extremity hemorrhages during care under fire. It is initially placed over clothing, high and tight. The deliberate application of a tourniquet is addressed when behind cover and during tactical field care to ensure proper hemorrhage control. The tourniquet is placed under clothing 2 to 3 inches above the wound. The application time is written on the tourniquet. Combat gauze is the hemostatic dressing of choice.

Airway — establish and maintain a patent airway. A second survivable cause of death at the POI is a non-patent (closed) airway. Airway injuries typically occur from maxillofacial trauma or inhalation burns. A conscious and speaking casualty has a patent open airway. An unconscious casualty who is breathing may benefit from the nasopharyngeal airway (NPA), in person medical training before the fact can assist with building this skillset.

Respiration — decompress suspected tension pneumothorax, seal open chest wounds, and support ventilation/oxygenation as required. The third potentially survivable cause of death on the battlefield is the development of a tension pneumothorax (PTX). Air trapped in the chest cavity begins to displace functional lung tissue and places pressure on the heart, resulting in cardiac arrest. Seal open chest wounds with a vented chest seal. A PTX is addressed via needle chest decompression (NCD) using a 14-gauge, 3.25-inch-long needle with a catheter. This is only for the professionally trained on your team, having several team members with professional, formal medical training and supplies to do so are highly desired. Chest seals are one thing, but decompression is done only if trained medical personnel are available to do so. This is NOT a task that basic first aid skills will qualify you to do.

Circulation — establish intravenous (IV) access and administer fluids as required to treat shock. Control of bleeding takes precedence over infusing fluids. Only individuals in shock or those who need intravenous (IV) medications need to have IV access established. Clinical signs of shock on the battlefield are: 1) unconsciousness or altered mental status not due to coexisting traumatic brain injury (TBI); and/or 2) abnormal radial pulse. Again, IV availability is subject to med supply and trained personnel among you or your team to do so; this is only for the professionally trained on your team.

Head injury/Hypothermia — prevent/treat hypotension and hypoxia to prevent worsening of traumatic brain injury and prevent/treat hypothermia.

Survival medicine requires us to understand these fundamentals of trauma first aid (MARCH) and survival medicine. Assess and prioritize applicable fundamentals from trauma first aid and survival medicine in a resource-constrained environment and execute those fundamentals under sub-optimal healing conditions. Just understand that medical personnel and facilities may rarely be available or may be overwhelmed in a widespread event.

Fundamentals of First Aid

When a nonmedical Citizen encounters an unconscious or injured individual, he must quickly and accurately evaluate the situation and the casualty to determine if it is safe for him to act as well as what, if any, first aid measures may be needed to prevent further injury or death. He should ask if trained medical personnel are available or direct someone else at the scene to call for or locate trained medical personnel. To prevent further injury to the casualty, once first aid has begun to be administered there should be no interruptions and those efforts should continue until such time as that person is relieved by medical personnel or the unit leadership directs them. You may also have to depend upon your own first aid knowledge and skills to save yourself (self-aid). A thorough understanding of the fundamentals of first aid can save a life, prevent permanent disability, or reduce long periods of hospitalization by knowing WHAT to do, WHAT NOT to do, and WHEN to do it.

The following key terms are identified and described in order to provide a further understanding of first aid. The key terms are presented in random order, not in order of importance.

Definitions

Casualty Evacuation - Nonmedical units use this to refer to the movement of casualties by foot or aboard nonmedical vehicles without en route medical care.

Combat Lifesaver - Combat lifesavers (a term borrowed from the Army) are nonmedical Citizens that have additional medical training beyond basic first aid / CPR. They are identified by their group for having that additional training beyond basic first aid procedures. Combat lifesavers (CLS) provide enhanced first aid for injuries. EMT-Bs (with some additional training) are a good starting point to assign CLS functions to in your teams.

Medic - Medics are the first individuals in the medical chain that make medically substantiated decisions based on medical certifications and specific training. Experienced EMTs, Paramedics, Fire Medics are all good examples of folks you want in your group to fill this role.

Emergency Medical Treatment - Emergency medical treatment is the immediate application of medical procedures to the wounded, injured, or sick by specially trained medical personnel. This will be unavailable / inaccessible in its current form if there is a widespread kinetic event.

Enhanced First Aid - Enhanced first aid is administered by the combat lifesaver. It includes measures which require an additional level of training above self-aid and buddy aid.

First Aid (Self and Buddy Aid) - Urgent and immediate lifesaving and other measures which can be performed for casualties (or performed by the casualty himself) by nonmedical personnel when medical personnel are not immediately available.

Medical Evacuation - Medical evacuation is the process of moving any person who is wounded, injured, or ill to and/or between medical treatment facilities while providing enroute medical care. Also referred to as MEDEVAC in relation to the 9-line medical evacuation request. This is a tough issue to address for a non-standard, irregular force that may not have robust friendly medical system support. Planning and wargaming how you will take care of the wounded must be done ahead of time and rehearsed during mission prep.

Medical Treatment - Medical treatment is the care and management of wounded, injured, or ill personnel by medically trained personnel.

Medical Treatment Facility - Medical treatment facility is any facility established for the purpose of providing medical treatment. This includes acute care clinics, ERs, dispensaries, and hospitals.

Tactical Combat Casualty Care - Tactical combat casualty care is often referred to as TC3. Tactical combat casualty care is prehospital care provided in a tactical-combat setting. Tactical combat casualty care is divided into the following three stages:

> Care under fire
> Tactical field care
> Tactical evacuation

Initial Encounter

Proper conduct at the initial encounter of the casualty coupled with appropriate movement and transport is important in the successful provision of first aid. Appropriate decisions and first aid task execution helps to determine the health and well-being of the casualty. When a casualty is first encountered it is imperative that the responder quickly and accurately assess what has occurred, determine the nature and extent of injuries and what (if any) first aid measures are appropriate and necessary. Accurately assessing the situation is as important for the safety and well-being of the responder as it is for the casualty. For example, if the casualty is being electrocuted the responder must not directly grab the casualty or he too will become a casualty.

Casualty Transport. Transporting or moving a casualty by the individual providing first aid must be carefully considered for a number of reasons. An example of this type of consideration may be based on the casualty having been involved in a motor vehicle crash. When responding to an accident, first aid providers must consider the possibility of injury to the casualty's spine before extracting the casualty from the vehicle. In this situation moving the casualty may be ill advised unless there is an immediate life-threatening situation such as fire, explosion where the casualty may be at risk of greater injury or death unless moved promptly. *Always* defer to medical system certified first responders for the movement of casualties in other than combat or extremely exigent circumstances (like a vehicle is on fire and the person will be burned alive unless moved).

If there is no danger of greater injury to the casualty by leaving them where they are found, first aid responders should render such aid as is necessary and wait for trained medical personnel to arrive. Once medical personnel are on site they can accurately treat the casualty and direct how and when they should be transported or moved.

Evaluate a Casualty. Evaluation of a casualty is necessary to identify and treat all life-threatening conditions and other serious wounds. Rapid and accurate evaluation of the casualty is the key to providing effective first aid.

Care Under Fire (CUF)

Return fire, use handheld smoke, direct suppressive fire at the threat to protect the casualty and yourself before providing first aid.

Determine if the casualty is alive or dead. In combat, the most likely threat to the casualty's life is from bleeding. Attempts to check for airway and breathing may unnecessarily expose the rescuer to enemy fire. It is an individual call you will have to make while attempting to provide first aid when your own life is in imminent danger. If you find a casualty with no signs of life, no pulse, no breathing, it is an individual choice to attempt to restore the airway or continue first aid measures while under fire. Just understand the risks of doing so.

Provide care to the living casualty. Direct the casualty to return fire, move to cover, and administer self-aid (stop bleeding), if they are able.

Reducing or eliminating enemy fire may be more important to the casualty's survival than the treatment you can provide. If the casualty is unable to move and you are unable to move the casualty to cover and the casualty is still under direct enemy fire, as a last resort simply have the casualty play dead to lessen the odds of being re-engaged.

Once enemy fire has been suppressed, in a battle-buddy team, approach the casualty (use smoke or other concealment if available) using the most direct route possible. Administer lifesaving hemorrhage control while determining the relative threat of enemy fire versus the risk of the casualty bleeding to death. If the casualty has severe, life-threatening bleeding from an extremity or has an amputation of an extremity, administer lifesaving hemorrhage control by applying a tourniquet from the casualty's IFAK before moving the casualty.

The only treatment that should be given at the point of injury is a tourniquet to control life threatening extremity bleeding. Move the casualty, his weapon, and mission-essential equipment when the tactical situation permits and recheck bleeding control measures (tourniquet) as soon as behind cover and not under enemy fire.

WARNING
If a broken neck or back is suspected, do not move the casualty unless to save his life.

Evacuation will be a challenge for guerilla forces. Non-standard evacuation must be planned for and practiced during training.

Tactical Field Care

Once under cover and not under hostile fire, perform tactical field care:

When evaluating and/or treating a casualty, seek medical aid as soon as possible. Do not stop first aid. If the situation allows, send another person to find medical aid.

Form a general impression of the casualty as you approach (extent of injuries, chance of survival). If a casualty is being burned, take steps to remove the casualty from the source of heat/fire before continuing evaluation and first aid.

Ask in a loud, but calm, voice: *Are you okay?* Gently shake or tap the casualty on the shoulder. Determine the level of consciousness by using the mnemonic AVPU: A = Alert; V = responds to Voice; P = responds to Pain; U = Unresponsive. To check a casualty's response to pain, rub the breastbone briskly with a knuckle or squeeze the first or second toe over the toenail. If casualty is wearing individual body armor, pinch his nose or his earlobe for responsiveness. If the casualty is conscious, ask where his body feels different than usual, or where it hurts. If the casualty is conscious but is choking and cannot talk, stop the evaluation and begin appropriate first aid.

Massive Bleeding	Identify uncontrolled bleeding and control with pressure or tourniquet.
Airway	Open airway by patient position or with airway adjuncts.
Respiration	Identify and seal open chest wounds with occlusive dressing.
Circulation	Start intravenous (IV) therapy, if needed (if available and trained to do so) and treat for shock
Head Injury / Hypothermia	Prevent/treat hypotension and hypoxia to prevent worsening of traumatic brain injury and prevent/treat hypothermia.

Bleeding. External, life-threatening bleeding. Bleeding is life threatening if any one of the following signs or symptoms are observed:

There is a traumatic amputation of an arm or leg.
There is pulsing or steady bleeding from the wound.
Blood is pooling on the ground.
The overlying clothes are soaked with blood.
Bandages or makeshift bandages used to cover the wound are ineffective and steadily becoming soaked with blood.
There was prior bleeding, and the casualty is now in shock (unconscious, confused, pale).

Arterial. Blood vessels called arteries carry oxygenated blood away from the heart and throughout the body. Arterial bleeding is the most serious type of bleeding. If not controlled promptly, it can be fatal. With this type of bleeding, the blood is typically bright red to yellowish in color and exits the wound in distinct spurts or pulses that correspond to the rhythm of the heartbeat rather than in a steady flow. Because the blood in the arteries is under high pressure, an individual can lose a large volume of blood in a short period.

Venous. Venous blood is blood that is returning to the heart through blood vessels called veins. A steady flow of dark red, maroon, or blackish in color blood characterizes bleeding from a vein due to the lack of oxygen it transports. Venous bleeding is still of concern. While the blood loss may not be arterial, it can still be quite substantial, and can occur with surprising speed without intervention. It can usually be controlled more easily than arterial bleeding.

Capillary. The capillaries are the extremely small vessels that connect the arteries with the veins. Capillary bleeding most commonly occurs in minor cuts and scrapes and generally oozes in small amounts as opposed to squirting (arterial) or flowing (venous). This type of bleeding is not difficult to control.

If the casualty has severe, life-threatening bleeding from an extremity or has an amputation of an extremity, administer life-saving hemorrhage control by applying a combat application tourniquet (CAT) or SOF-T from the casualty's IFAK before moving the casualty. Personnel with life-threatening bleeding can bleed to death from a complete femoral artery and vein disruption within as little as three minutes. Control life-threatening bleeding immediately because replacement fluids are most likely not available.

If a commercial tourniquet is unavailable, apply an improvised tourniquet made from a rod (made from a jack handle, stick, scabbard, cleaning rod, pipe, dowel), a band of material at least 1-1/2 inches wide (made from a cravat, bandana, towel, ace bandage, shirt, nylon webbing, rifle sling.

Belts, zip ties, should only be used as a last resort, and a securing mechanism as a constricting or compressing device to control arterial and venous blood flow to a damaged extremity for a short period of time.

The wide band of material is made into a loop that fits over the damaged limb, 2-3 inches above the site of arterial bleeding, and tied tightly with an overhand knot. Next, lay the rod across the overhand knot.

The running ends of the loop are then used to tie another overhand knot on top of the rod forming a square knot with the rod through the center of the knot. The rod is then twisted, applying pressure around the limb tight enough to stop the arterial bleeding. Do not tighten the tourniquet more than necessary to stop the bleeding. In the case of amputation, dark oozing blood may continue for a short time. This is the blood trapped in the area between the wound and tourniquet. Fasten the tourniquet to the limb by looping the free ends of the tourniquet over the ends of the stick. Then bring the ends around the limb to prevent the stick from loosening. Tie them together on the side of the limb.

If bleeding continues, place a second tourniquet side by side to the first tourniquet. A tourniquet can be left in place up to two hours without damage to vessels, nerves, muscle or loss of limb. If isolated, the victim or buddy should release the pressure from the tourniquet after two hours, and then retighten if blood loss continues.

Hypoxia is the result of insufficient oxygen in the blood. It is a potentially deadly condition and one of the leading causes of cardiac arrest. Cardiac arrest is linked to an absence of circulation in the body, for any one of a number of reasons. For this reason, maintaining circulation is vital to moving oxygen to the tissues and carbon dioxide out of the body.

Check to see if the casualty has a partial or complete airway obstruction. If they can cough or speak, allow them to clear the obstruction naturally. Stand by, reassure the casualty, and be ready to clear their airway and perform mouth-to-mouth resuscitation should they become unconscious.

Open the Airway by placing the casualty in the recovery position by rolling him as a single unit onto his side, placing the hand of his upper arm under his chin, and flexing his upper leg.

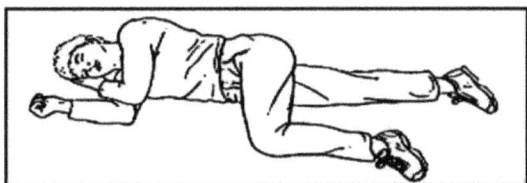

Recovery Position

Watch the casualty closely for life-threatening conditions and check for other injuries, if necessary. If the casualty is not breathing, immediately seek medical aid.

If the casualty resumes breathing at any time during this procedure, the airway should be kept open and the casualty should be monitored. If the casualty continues to breathe, he should be transported to medical aid or medical treatment facility in accordance with the tactical situation. If their airway is completely obstructed, administer abdominal thrusts until the obstruction is cleared. Any one of the following can cause airway obstruction, resulting in stopped breathing:

Foreign matter in mouth of throat that obstructs the opening to the trachea.

Face or neck injuries.

Inflammation and swelling of mouth and throat caused by inhaling smoke, flames, and irritating vapors or by an allergic reaction.

"Kink" in the throat (caused by the neck bent forward so that the chin rests upon the chest).

Tongue blocks passage of air to the lungs upon unconsciousness. When an individual is unconscious, the muscles of the lower jaw and tongue relax as the neck drops forward, causing the lower jaw to sag and the tongue to drop back and block the passage of air.

Using a finger, quickly sweep the casualty's mouth clear of any foreign objects, broken teeth, dirt etc

Using the jaw thrust method (figure below), grasp the angles of the casualty's lower jaw and lift with both hands, one on each side, moving the jaw forward. For stability, rest your elbows on the surface on which the casualty is lying. Use the index fingers to push the angles of the casualty's lower jaw

* Grasp the angles of the lower jaw and lift with both hands, one on each side, moving the jaw forward.

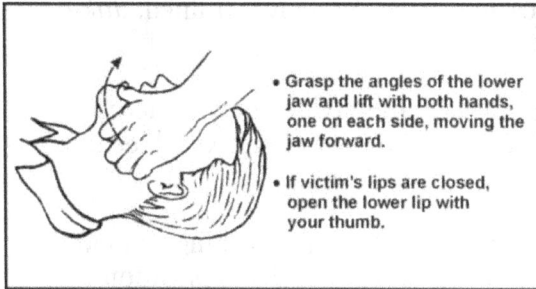

* If victim's lips are closed, open the lower lip with your thumb.

Jaw Thrust Method
Clearing an Airway

forward. If the casualty's lips are still closed after the jaw has been moved forward, use your thumbs to retract the lower lip and allow air to enter the casualty's mouth.

With the casualty's airway open, pinch their nose closed with your thumb and forefinger and blow two complete breaths into their lungs. Allow the lungs to deflate after the second inflation and:

Look for the chest to rise and fall.
Listen for escaping air during exhalation.
Feel for flow of air on your cheek.

If the forced breaths do not stimulate spontaneous breathing, maintain the casualty is breathing by performing mouth-to-mouth resuscitation (rescue breathing). There is danger of the victim vomiting during mouth-to-mouth resuscitation. Check the victim's mouth periodically for vomit and clear as needed.

If the casualty is unconscious, if respiratory rate is less than 2 in 15 seconds, and/or if the casualty is making snoring or gurgling sounds, insert a nasopharyngeal airway (NPA) from the casualty's IFAK. (only if trained...*take a class!*)

Cardiopulmonary resuscitation (CPR) may be necessary after cleaning the airway, but only after major bleeding is under control.

Continue to check for bleeding by performing a blood sweep. Control external bleeding by the application of direct pressure, indirect pressure, elevation, or digital ligation.

Direct Pressure. The most effective way to control external bleeding is by applying pressure directly over the wound. This pressure must not only be firm enough to stop the bleeding, but it must also be maintained long enough to "seal off" the damaged surface. If bleeding continues after having applied direct pressure for 30 minutes, apply a pressure dressing. This dressing consists of a thick dressing of gauze or other suitable material applied directly over the wound and held in place with a tightly wrapped bandage. These types will include the Israeli dressing or even the old school battle dressings. It should be tighter than an ordinary compression bandage but not so tight that it impairs circulation to the rest of the limb (*take some in person classes!*). Once you apply the dressing, do not remove it, even if it becomes blood soaked. Leave the pressure dressing in place for 1 or 2 days, after which you can remove it and replace it with a smaller dressing. In a long-term survival environment, make fresh, daily dressing changes and inspect for signs of infection.

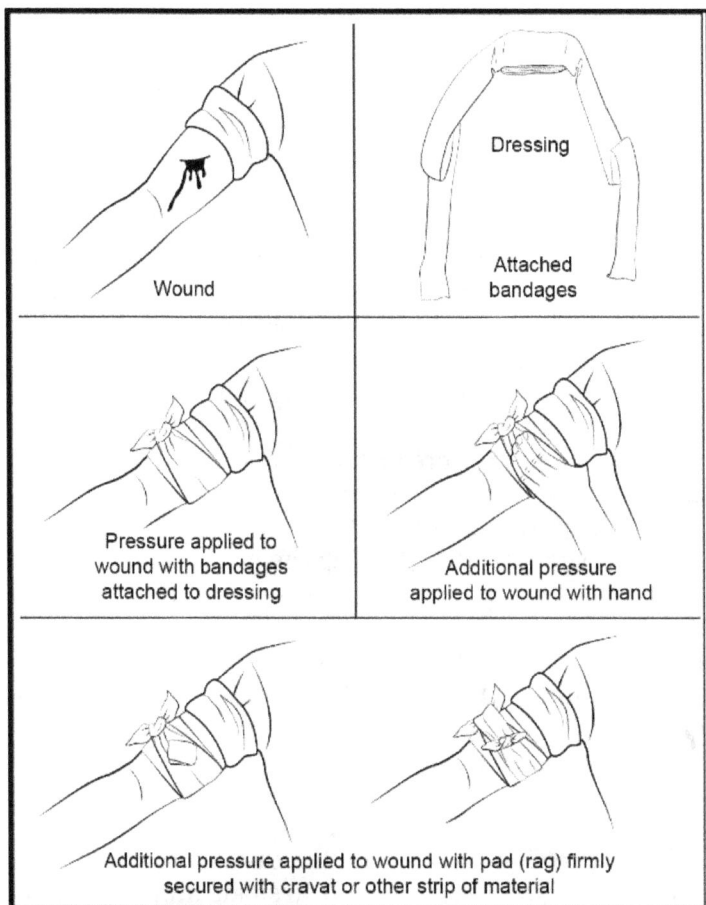

Wound

Dressing

Attached bandages

Pressure applied to wound with bandages attached to dressing

Additional pressure applied to wound with hand

Additional pressure applied to wound with pad (rag) firmly secured with cravat or other strip of material

The purpose of a dressing is to control bleeding, absorb wound secretions, and to prevent bacteria from entering the wound. Materials that make functional field expedient dressings include cloth from a shirt, undergarments, socks, bandanas, handkerchiefs, thin towels etc. Cut these materials to proper size to cover the wound and sterilized before use. To sterilize, place the material in water for ten full minutes at a rolling boil. If needed, clean and sanitize used bandages by boiling them, then reuse them if no other option exists. This is less than ideal...but the systems may not be present to support what we consider ideal.

Raising an injured extremity as high as possible above the heart's level slows blood loss by aiding the return of blood to the heart and lowering the blood pressure at the wound. However, elevation alone will not completely control bleeding; apply direct pressure over the wound.

Breathing and chest injuries. Fractured ribs are common, painful, and disabling. The isolated person will not have access to pain medications and must understand that the pain associated with rib injuries can lead to reduced movement and cough suppression which can contribute to formation of secondary chest infection. To treat fractured ribs:

Protect the injured rib by supporting the arm on the injured side with a sling-and-swathe.

Encourage the person to take deep breaths regularly, even if it hurts, to keep the lungs clear.

Watch the person for increasing trouble breathing.

Prevent or Control Shock. Check the casualty for signs and symptoms of shock. Symptoms may include sweaty but cool skin, pale skin, restlessness or nervousness, thirst, severe bleeding, confusion, rapid breathing, blotchy blue skin, nausea and/or vomiting.

Position the Casualty. Move the casualty under a permanent or improvised shelter to shade him from direct sunlight. Lay the casualty on his back unless a sitting position will allow the casualty to breathe easier and elevate the casualty's feet higher than the heart using a stable object so the feet will not fall. Loosen clothing at the neck, waist, or anywhere it is binding and prevent the casualty from getting chilled or overheated. Using a blanket or clothing, cover the casualty to avoid loss of body heat by wrapping completely around the casualty. Ensure no part of the casualty is touching the ground, as this increases loss of body heat.

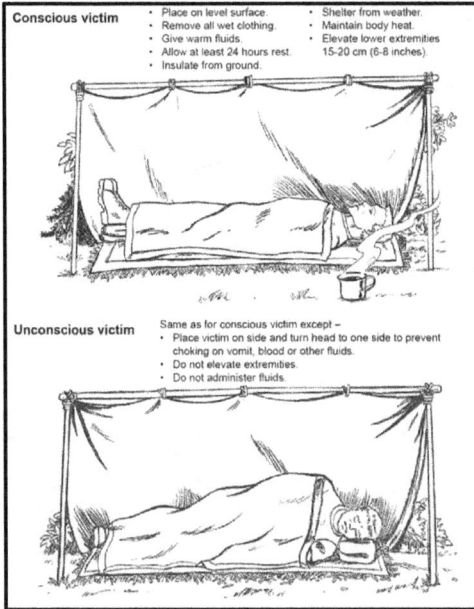

Conscious victim	• Place on level surface. • Remove all wet clothing. • Give warm fluids. • Allow at least 24 hours rest. • Insulate from ground.	• Shelter from weather. • Maintain body heat. • Elevate lower extremities 15-20 cm (6-8 inches).
Unconscious victim	Same as for conscious victim except – • Place victim on side and turn head to one side to prevent choking on vomit, blood or other fluids. • Do not elevate extremities. • Do not administer fluids.	

Calm and Reassure the Casualty by taking charge and showing self-confidence. You should assure the casualty that he is being taken care of and watch them closely for life-threatening conditions and check for other injuries, if necessary.

No matter the injury, whether superficial or severe, you cannot react negatively or show any emotion or doubt in front of the injured. Saying "man that is bad" or alluding to the extent can and will have a negative affect on your casualty or patient. Go into work mode and take care of your buddy.

These are just a few examples of medical issues that may be encountered by the pro Citizen. The recurring "take a class" comments in this section are not meant to be funny, dismissive, or condescending. We want to constantly reinforce that reading or watching content are not adequate means to prepare one to perform first aid or life saving measures.

Weather and Environmental Related Injuries

Hot and cold weather injuries can be mild to life threatening. Environmental injuries range from bites and stings, to accidental exposure to poisonous plants. Knowing the signs and symptoms, and the proper treatment is crucial.

HEAT INJURIES		
Injury	Signs and Symptoms	First Aid
Heat Cramp	Casualty experiences muscle cramps in arms, legs, or stomach. May also have wet skin and extreme thirst.	1. Move the casualty to a shaded area and loosen clothing. 2. Allow casualty to drink one quart of cool water slowly every hour. 3. Monitor casualty and provide water as needed. 4. Seek medical attention if cramps persist.
Heat Exhaustion	Casualty experiences loss of appetite, headache, excessive sweating, weakness or faintness, dizziness, nausea, or muscle cramps. The skin is moist, pale, and clammy.	1. Move the casualty to a cool, shaded area and loosen clothing. 2. Pour water on casualty and fan to increase cooling effect of evaporation. 3. Provide at least one quart of water to replace lost fluids. 4. Elevate legs. 5. Seek medical aid.
Heat Stroke (Sunstroke)	Casualty stops sweating (hot, dry skin), may experience headache, dizziness, nausea, vomiting, rapid pulse and respiration, seizures, mental confusion. Casualty may suddenly collapse and lose consciousness.	1. Move casualty to a cool, shaded area, loosen clothing, and remove outer clothing if the situation permits. 2. Immerse in cool water. If cool bath is not available, pour cool water on the head and body. Fan casualty to increase the cooling effect of evaporation. 3. If conscious, slowly consume one quart of water. HEAT STROKE IS A MEDICAL EMERGENCY! Evacuate and get to definitive care

COLD INJURIES		
Injury	Signs and Symptoms	First Aid
Chilblain	Red, swollen, hot, tender, itchy skin. Continued exposure may lead to infected (bleeding, ulcerated) skin lesions.	1. Area usually responds to locally applied warming (body heat). 2. Do NOT rub or massage area. 3. Seek medical treatment.
Immersion (Trench) Foot	Affected parts are cold and numb. As body parts warm, they may become hot, with burning and shooting pains. Advanced stage: Skin is pale with bluish cast; pulse decreases; blistering, and swelling occur. Swelling, heat hemorrhages, and gangrene may follow.	1. Gradual warming by exposure to warm air. 2. Do NOT massage or moisten skin. 3. Protect affected parts from trauma. 4. Dry feet thoroughly and avoid walking. 5. Seek medical treatment.
Frostbite	Superficial: Redness, blisters in 24-to-36 hours followed by peeling skin. Deep: Preceded by superficial frostbite: skin is painless, pale-yellowish, waxy, "wooden" or solid to the touch, blisters form in 12-to-36 hours.	Superficial: 1. Keep casualty warm and gently warm affected parts. 2. Decrease constricting clothing, increase exercise and insulation. Deep: 1. Protect the part from additional injury. 2. Seek medical treatment as fast as possible.
Snow Blindness	Red, scratchy, or watery eyes; headache; increased pain in eyes with exposure to light.	1. Cover the eyes with a dark cloth. 2. Seek medical treatment.
Dehydration	Similar to heat exhaustion.	1. Keep warm and loosen clothes. 2. Replace lost fluids, rest, and seek additional medical treatment.
Hypothermia	Casualty is cold, shivers uncontrollably (jack hammering) until shivering stops. Can affect consciousness. Uncoordinated movements, shock, and coma may occur as body temperature drops.	Mild hypothermia: 1. Warm body evenly and without delay. (Provide a heat source.) 2. Keep dry, protect from elements. 3. Warm liquids may be given only to a conscious casualty. 4. Be prepared to start CPR. 5. Seek medical treatment immediately. Severe hypothermia: 1. Quickly stabilize body temperature. 2. Attempt to prevent further heat loss. 3. Handle the casualty gently. 4. Evacuate to nearest medical treatment facility as soon as possible.

ENVIRONMENTAL INJURIES

Injury	First Aid
Snake bite	1. Get the casualty away from the snake. 2. Remove all rings and bracelets from the affected extremity. 3. Reassure the casualty and keep them quiet. 4. Apply constricting band(s) two-to-three inches above the bite. 5. Immobilize the affected limb below the level of the heart. 6. Treat for shock and monitor. 7. Kill the snake (without damaging its head or endangering yourself, if possible) and send it with the casualty. 8. Evacuate and seek medical treatment immediately.
Brown Recluse or Black Widow spider bite	1. Keep the casualty calm. 2. Wash the area. 3. Apply ice or a freeze pack, if available. 4. Seek medical treatment.
Tarantula bite, scorpion sting, or ant bite	1. Wash the area. 2. If site of bite(s) or sting(s) is on the face, neck (possible airway blockage), or genital area, or if the reaction is severe (or it was a dangerous southwestern scorpion sting), keep the casualty as quiet as possible, administer an antidote, if needed, and seek immediate medical aid.
Wasp or bee sting	1. If the stinger is present, remove by scraping with a knife or fingernail. Do NOT squeeze venom sack on stinger, more venom may be injected. 2. Wash the area. 3. Apply ice or freeze pack, if available. 4. If allergic signs or symptoms appear, be prepared to administer an antidote and seek medical assistance.
Human or animal bites	1. Cleanse the wound thoroughly with soap or detergent solution. 2. Flush bite well with water. 3. Cover bite with a sterile dressing. 4. Immobilize injured extremity. 5. Transport casualty to a medical treatment facility. 6. For human bites, try to extract some of the attacker's saliva from the wound and send that in a sealed, identified container with the casualty. (totally dependent on the status of the medical treatment system) 7. For animal bites, kill the animal without endangering yourself or damaging the animal's head, and send its head with the casualty. (totally dependent on the status of the medical treatment system)
Poison ivy, oak, or sumac	1. Gently clean affected area two-to-three times daily. Wash clothing. 2. Apply topical anti-itch lotion or ointment as needed, and cover. 3. Avoid scratching the area. 4. Observe for signs of infection (increasing redness, tenderness, warm to the touch). 5. Seek medical attention, if needed.

The Professional Citizen IFAK

Your IFAK should be accessible and clearly marked. Unit SOPs must dictate how they are marked and where they are placed so a teammate can locate a casualty's IFAK. Remember you will use the casualty's med gear first, not your own (unless exigent circumstances require of course, we wouldn't deny care because our battle buddy's IFAK was consumed or lost).

Contents List. This is a baseline or *minimum* list, addition of other items is encouraged. Airways (if trained) etc are always great to have on hand.

> (1) Tourniquet (CAT or SOF-T) (additional TQs should be carried on gear or on person)
> (1) Emergency Trauma /Israeli Dressing - 4"
> (1) Chest Seal (Vented)
> (2) Wound Packing Gauze
> (2) Gloves (not black, blood will not show well on black gloves during blood sweeps)
> (1) Duct Tape-Mini - 2" X 100"
> (1) Casualty Card

Three hours *without shelter in killer weather*

Shelter

A shelter can protect you from the sun, insects, wind, rain, snow, hot or cold temperatures, and even enemy observation. In some areas, your need for shelter may take precedence over your need for food and possibly even your need for water (rule of threes). For example, prolonged exposure to cold can cause excessive fatigue and weakness and sun exposure can result in life threatening heat injuries.

Your primary shelter will be your clothing or uniform. This point is true regardless of whether you are in a hot, cold, tropical, desert, or frigid situation. For your uniform to protect you, it must be in as good a condition as possible and must be worn properly.

Fabric. One of the most important considerations is the fabric of your clothing. In temperate regions, you need clothing that can keep you warm in cold weather and cool in warm weather. Synthetic fabrics like polyester and nylon are good options because they are quick-drying and lightweight, but still provide insulation. Synthetics can melt however, so be cautious matching a full synthetic article of clothing with an environment where burn or flame injuries are commonplace. Avoid 100 percent cotton in most environments, as it retains moisture and can get very heavy when wet making it a poor choice for hard use.

Layering. To adapt to the changing weather conditions, it's important to have a layering system in place for cooler weather. A base layer should be made of moisture-wicking fabric to keep you dry and comfortable. The higher end merino wool manufacturers have broken the code on wool, the modern versions are comfortable, anti microbial and will stay (relatively) warm even when wet. A mid-layer should provide insulation, and an outer layer should be wind and waterproof. Make sure to choose layers that can be removed or added as needed to maintain a comfortable temperature.

The "COLDER" (clean, overheating, loose layers, dry, examine, and repair) principle that the Army uses is a decent guide and provides a foundation for care and wear of outdoor clothing or uniform while in an unsupported environment. It is further described in the following:

Clean. Keep clothing clean. This principle is always important for sanitation and comfort. In winter, it is also important from the standpoint of warmth. Clothes matted with dirt and grease lose much of their insulation value. Heat can escape more easily from the body through the clothing's crushed or filled up air pockets. Dirt within the fibers will cut or tear the clothing, causing them to wear out prematurely. In an unsupported situation, it may be impractical to wash clothing often; therefore, take the as much care as you can to prevent clothing from becoming soiled.

Overheating. Avoid overheating. When you get too hot, you sweat and clothing absorbs moisture. This affects your warmth as dampness decreases the insulation quality of clothing and as sweat evaporates, the body cools. Adjust your clothing so that you do not sweat, start cold when conducting movements. A cold start will help mitigate sweating. Do not overdress when you anticipate movement, save warmer clothing for stationary times / pulling security. Keep the human heat engine at bay by partially opening your jacket or shell, using pit zips, removing an inner layer of clothing, removing heavy outer mittens, or by throwing back your parka hood or changing to lighter headgear. The head and hands act as efficient heat dissipaters when overheated.

Loose layers. Wear clothing loose and in layers. Tight clothing and footgear restrict blood circulation and invite cold injuries. It also decreases the volume of air trapped between the layers, reducing the insulation value. Several layers of lightweight clothing are better than a single, equally thick layer of clothing, because the layers have dead airspaces between them. This dead airspace provides extra insulation. In addition, layers of clothing allows you to take off or add layers to prevent excessive sweating or to increase warmth.

Dry. Keep clothing dry. This is important since wet clothing can conduct heat away from the body up to 25 times faster than dry clothing, depending on the type of fabric. Being cold is mostly tolerable, being wet in warm temps is simply uncomfortable. Being cold *and* wet is a recipe for disaster. It has nothing to do with "being tough", hypothermia is a killer that will creep up and can turn a unit combat ineffective. Wear water repellant clothing as an outer layer if available and the temperature supports doing so (avoid overheating), it will shed most of the water from rain and snow. Despite the precautions taken, there will be times when unsupported patrols cannot avoid getting wet. Getting into dry clothing is a priority. As you move about, you can place wet clothing on your rucksack exposing the wet clothing to the sun and wind; this will dry them. They can also put damp clothing near their body's core and body heat will began to dry the clothing. If at a stationary location, construct drying racks or a clothesline to hang clothes up to dry. Having dry socks in your ruck to change in to at night is a lifesaver. Dry leather items slowly. If no other means are available for drying boots you can put the boots between their sleeping bag liner and shell if you are using one when not pulling security. Your body heat will help dry the leather. This has to be weighed against the risk of reaction time should you have contact. Having dry boots may outweigh the requirement for this at some point. METT-TC applies as always.

Examine. Periodically examine clothing for worn areas, tears, and cleanliness.

Repair. Repair any damaged clothing before tears and holes become too large to patch or sew. Patches can be made from cloth, tape, and other materials. Use a sewing kit, leaders should consider having one small gear repair kit carried among the patrol when outside the wire.

SOPs will dictate what is acceptable for individual shelter when use is appropriate during patrols; in woodland environments these will usually consist of ground hugging poncho or camouflage tarp shelters that are *not taller than waist level, knee high is ideal*. It isn't a camp out or a survival outing. The purpose of being out is to conduct missions against a thinking enemy; stealth and signature mitigation always take precedent. During training if the occasion arises when you must construct a poncho shelter or use a low profile bivy take the opportunity to walk out and observe it from the threat perspective. How far away can you see it? Did you blend in with the surroundings and break up the shape by using vegetation and the terrain? How long will it take you to shove it in your ruck and leave the area? If you are not pulling security can you be out of it and in the fight quickly? All things that must be rehearsed on an individual basis so we don't become "that guy" in our groups...or worse.

Shelter requirements may include thermal camouflage. We won't dive too deep on thermal camo in this particular manual, but be aware that any thermal camo (most likely a suspended casualty blanket) must remain at least 12 inches off of your body to prevent your heat signature from bleeding through.

The acronym BLISS is another acronym to help guide your shelter selection and setup. The acronym BLISS stands for:

B - Blend in with the surroundings. Choose a shelter option that will not stand out in your surroundings. Match camouflage patterns and colors to your AO and season.

L - Low silhouette. Keep the highest point below waist level. Knee high is much better, but waist or below is a good standard for general purposes.

I - Irregular shape. Lines, especially horizontal lines, are easily detected by the human eye. A horizontal straight line in nature is a recognizable indicator of presence.

S - Small. Having the mansion hooch is a no-go. Build or select a solution that is just large enough. This not only helps with the visual signature, it will assist in keeping your carried load volume and weight down.

S - Secluded location. Patrol base selection is a leader task, leaders must select areas that are defensible and not easily detected. If you are alone be sure to choose areas away from human traffic, both current and potential. Staying away from those natural lines of drift we discussed in the Land Nav section will help minimize the chances of detection.

Commercial or military one man shelters. One person bivy or single pseudo-tents are a solution that has to be vetted and tried out in all seasons and samples of bad weather. These shelters can range in quality and function from excellent to just being leaky caskets. Weight, durability, speed and ease of setup and teardown, and exit speed all must be assessed before you choose one of these as a solution.

Poncho Lean-tos. It takes only a short time and minimal equipment to build this lean-to. For the lean-to, required items are a poncho, 7 to 10 feet of 550 or comparable line, three stakes (carried or made), and two trees 7 to 10 feet apart. Before selecting the trees to be used or the location of the poles, check the wind direction. Ensure that the back of the lean-to will be into the wind. Use the following steps to construct a poncho lean-to:

Step 1. Tie off the hood of the poncho. Pull the drawstring tight, roll the hood long ways, fold it into thirds, and tie it off with the drawstring.

Step 2. Cut the rope in half. On one long side of the poncho, tie half of the rope to the corner grommet. Tie the other half to the other corner grommet (pre-rig your preferred shelter method, you should not have to cut cord or affix anchor points in the field).

Step 3. Attach a drip stick (about a 4-inch stick) to each rope about 1 inch from the grommet. These drip sticks will keep rainwater from running down the ropes into the lean-to. Tying strings (about 4 inches long) to each grommet along the poncho's top edge will allow the water to run to and down the line without dripping into the shelter.

Step 4. Tie the ropes about waist high on the trees. Use a round turn and two half hitches with a quick-release knot.

Step 5. Spread the poncho and anchor it to the ground, putting sharpened sticks through the grommets and into the ground.

Step 6. If the lean-to will be used for more than one night or rain is expected, make a center support for the lean-to. Make this support with a line. Attach one end of the line to the poncho hood and the other end to an overhanging branch. Make sure there is no slack in the line. Alternatively, place a stick upright under the center of the lean-to. However, this method will restrict space and movements in the shelter.

Step 7. For additional protection from wind and rain, place some brush, a rucksack, or other equipment at the sides of the lean-to.

Step 8. To reduce heat loss to the ground, place some type of insulating material, such as an insulated mat, or field expedient method such as dry leaves or pine needles inside the lean-to. When at rest, you can lose as much as 80 percent of your body heat to the ground.

Step 9. To increase security from enemy observation, lower the lean-to's silhouette by making two changes. First, secure the support lines to the trees at knee height (not at waist height) using two knee-high sticks in the two center grommets (sides of lean-to). Second, angle the poncho to the ground, securing it with sharpened sticks, as above.

Poncho Tent. This tent provides a low silhouette. It also protects you from the elements on two sides. However, it has less usable space and observation area than a lean-to, decreasing reaction time to enemy detection. To make this tent, use a poncho, two 5- to 8- foot long 550 cord, six stakes, and two trees 7 to 10 feet apart. To construct a poncho tent, use the following steps:

Step 1. Tie off the poncho hood in the same way as the poncho lean-to.

Step 2. Tie a 5- to 8-foot rope to the center grommet on each side of the poncho.

Step 3. Tie the other ends of these ropes at about knee height to two trees 7 to 10 feet apart and stretch the poncho tight.

Step 4. Draw one side of the poncho tight and secure it to the ground pushing the stakes through the pre rigged loops on the grommets.

Step 5. Follow the same procedure on the other side.

Step 6. If a center support is needed, use the same methods as for the poncho lean-to. Another center support is an A-frame set outside but over the center of the tent. Use two 12- to 16-foot long sticks, one with a forked end, to form the A-frame. Tie the hood's drawstring to the A-frame to support the center of the tent.

Three days without water

Water resupply is difficult even for organized forces. Water is heavy (8.34lbs per gallon) and requires sterile containers for potable water transport. There are several general rules that apply to water and hydration. We must assume that we will not have a reliable resupply of water and plan accordingly. Running out of water in the field is an awful experience and can lead to widespread heat injury in a team. A multi-tiered approach when planning is necessary; counting on a water source enroute can be a primary means of resupply but the enemy and environment always get a vote (METT-TC). Water sources may become inaccessible due to terrain and weather or presence of enemy. If water is limited in your AO count on all the known watering holes being primed for ambush. It is also not beyond reason for the enemy to poison or contaminate water sources to prevent local resupply.

At a temperature of 68° F with limited physical activity, personnel will normally require 2 to 3 quarts of water a day to maintain efficiency. If an individual is thirsty they are already behind the power curve and are headed for dehydration. Water is necessary to replace what is lost through daily functions. Individuals lose about 1.4 quarts of water through urine loss, 1.0 quart of water through sweat, and 0.2 quarts of water through fecal matter per day. Water loss increases with heat exposure. When exposed to high temperatures, water loss from sweat increases to as much as 3.5 quarts per hour. At this rate, body fluids are quickly depleted. Physical activity increases loss of water in two ways: increased respiration and sweating due to excessive body heat. The time and effort required obtaining water and the decrease in the thirst response in cold weather also favors the development of dehydration. Dehydration in cold environments can be hidden and will sneak up on you. Drinking water in the cold is a matter of discipline and awareness. Individuals can also lose water through vomiting and diarrhea if they are ill with flu or intestinal issues.

Potable water is not necessarily palatable. Water that is safe to drink may not be appealing while water that is unsafe may "seem" fine. Palatability is NOT an indicator of potability.

If you must choose between running water and stagnant water, choose running water but make sure it is also appropriately treated.

Do not eat snow. The body has to heat the water and melt the snow once you eat it and is a foolish expenditure of calories that are already in short supply.

Do not drink saltwater, even a small amount.

Do not drink water found in natural depressions. It is stagnant water and is often heavily contaminated with pesticides, garbage and other debris.

Do not drink urine or alcoholic beverages.

Consider all surface water to be contaminated with human and animal waste with the possible exception of very high mountain streams or springs found in uninhabited areas. Even in these cases animals may have contaminated the source, so caution is always advised.

Procure water using a variety of different methods depending upon the source. The first step to procuring water is obtaining a container that can temporarily hold water to enable purification and that can be sealed for long-term storage.

Some examples of water container dry weights. Top row: Heavy Cover Titanium 1 liters (cup/no cup), Nalgene stainless 32 oz, Swiss 800ml canteen with cup. Middle row: Nalgene 1 qt canteens, US surplus 1qt canteens. Bottom row: CNOC 2l bladder, Hydrapak 32oz, smart water 1l bottle.

Water containers (canteens) in a cold weather environment should be placed upside down (if you are stationary) to enable its use when frozen. Since water freezes from top to bottom, any unfrozen water would be on the bottom where it is accessible. If the container is stored improperly ice will block the opening making the water inaccessible. Only use containers proven to withstand freezing in cold temperatures. Some plastic / composite containers handle freezing contents, and some do not. Having dual function containers that can be used to boil for disinfecting water are an excellent choice for unsupported operations. Some materials are better than others for this purpose, for example Titanium will be lighter and will cool much faster than steel. The upfront cost of titanium is significant, but the weight savings may be worth the investment for you. The 1L Smartwater™ bottles are great and will also fit a Sawyer filter. They are incredibly durable, cheap, and lightweight. Remember everything in this tactical and outdoor world is a tradeoff. The dry weights for some examples are in the figure, each has its advantages and disadvantages for patrol use. Combinations are the way to go in our opinion. A Camelbak™ bladder coupled with a metal container in your system is a good option. A mix like this provides the capability to process water with heat (metal containers) or chemical treatment (metal or camelback bladder) and has components available that can withstand freezing temperatures.

Processing water by boiling is an excellent method, however the use of an open fire is a dead giveaway in the field. Wood or debris fires should never be used when there is any chance of a nearby threat. Canister, solid fuel (which are lackluster at best) or alcohol stoves are all options but must be used sparingly and cautiously. Some stoves are loud, some give off strong odors, and some require bulky liquid fuel. All types will have a visible flame, so using after EENT is ill-advised. Filtering is a good method of purification but can be time-consuming for the quantities needed. Chemical methods using commercially available tablets are another viable means, but again will take time and you are limited by the container size. When using water purification tablets always follow the manufacturer's instructions for the recommended use of the product. Always verify that the tablets are within their specified stale date to ensure proper dosage. When using chemicals, you must remember to let the water set for the prescribed amount of time (usually colder or cloudy water takes longer, check the directions carefully). Shortly after dissolving the tablet or liquid treatment open the cap and let the treated water rinse any unpurified water from the threads, reseal for the remainder of the prescribed processing time to complete treatment. This prevents contamination of the threads before drinking.

Remote operations must always take into consideration the safety of various sources and procure water from the safest source. Some sources such as plants and certain types of precipitation do not require purification and make a great choice for personnel who do not have the ability to purify water. The volume (quantity) of water available will be far less than what is required, especially for a team that is on the move. Other sources can provide large quantities of water on demand but will require purification. Avoid water that has a foul odor or smell, and water potentially contaminated with toxins, faster moving water contains more oxygen and typically fewer harmful microorganisms than stagnant water and is usually a better option to procure water.

> The key is to have TTPs in place to minimize the burden and distraction of shelter, water, and food procurement for a unit in combat; don't make basic survival *the* mission.

Three weeks without food

Although you can live for several weeks without food, you need an adequate amount to stay physically capable and healthy. Without food your mental and physical capabilities will deteriorate rapidly, you will become weak and far less combat effective. Choosing calorie dense, compact, ready to eat calories takes some deliberate personal experimentation and application under extended field conditions. Commercial MRE's, backpacker meals, and regular old high calorie dry/portable food are all choices you will have to make. Stay away from canned/liquid as well as anything that requires cooking prep. Foods that generate an inordinate amount of trash, smells, or will be a pain to manage in the field will not be in the patrol's best interest. Other than these general guidelines and avoiding "surplus" MREs (we do not recommend buying these, you don't know what conditions they have been stored in) and empty / low calorie items the choices are wide open for experimentation. Convenience, calorie density, weight, SLLS, and food safety (spoilage) all must be considered. Using water that may be in short supply to prep a meal is probably a non-starter for you as well; just keep all this in mind as you try different options. This topic is worthy of an hour of class time for your group, or at a minimum worth asking other experienced members what works best for them.

The average person needs 2,000 calories per day to function at a minimum level. An average person is not one engaging in patrolling, rucking, and even fighting; that calorie requirement can rise to 4,500 or more (many more) calories burned per day. An adequate amount of carbohydrates, fats, and proteins without an adequate caloric intake will lead to starvation and cannibalism of the body's own tissue for energy. Signs of malnutrition include loss of body fat, difficulty breathing, reduced muscle mass, fatigue, and longer healing time for wounds and illness.

Operational environments play a significant role in determining how much food an individual requires. In cold environments, more calories will be required to maintain necessary body heat. In hot environments, more water and less food will be required to meet their needs. You should learn what and how to consume and ration your food and water. If you are low on water (less than one quart a day), they should stay away from starchy, dry, salty food sources. These types of items will require more water to digest and increase thirst. Consuming foods with high carbohydrate content is best for water conservation, while high protein foods tend to require more water for digestion.

Food procurement (living off the land, scavenging, hunting) are time consuming and should not be focused on as a solution in this early phase of your training. Tactical bushcraft and survival are critical skills that are addressed in detail later in the Pro Citizen reference series.

Fratricide Prevention

Units have other means of designating friendly forces from the enemy. Typically, these marking systems are derived from the unit tactical standard operating procedure (TACSOP) or other standardization publications, and applied to the personnel, small units, or vehicles as required.

Situational awareness and control measures. The most important and effective means of fratricide prevention is Situational Awareness (SA). This means that an individual or an organization is situationally aware. In simple terms it just means you know where you and friendly elements are. It will ebb and flow during maneuver, but you should always have an idea of where your teammates are, and where they should not be. There are additional ways to add to this awareness or mitigate the risk of fratricide using markings or signals (discussed below). Control measures and signals put in place during planning and by SOP will also help ensure you or your team does not cross into a unit boundary, or stumble into a friendly ambush or engagement area.

Markings. Unit markings are defined within the unit SOP. They distinctly identify an individual or vehicle as friendly in a standardized manner. We have seen examples from recent conflicts with vehicles painted with large letters or colored tape bands on the arms of combatants. Flags and banners have been used as well. All of these methods can be easily replicated and "stolen" by the opposing force or bad actors. Secondary means are always needed to ensure we do not engage friendlies and to prevent the enemy from slipping in and gaining a tactical advantage over our team.

Panels. VS-17 panels provide a bright recognition feature that allows to identify friendly vehicles through the day sight during unlimited visibility. While visible and recognizable the VS17 is in wide use and could be easily confused for a friendly, threat, or neutral individual. Use sparingly based on METT-TC assessments in your area.

Lighting. Chemical or light emitting diode lights provide a means of marking vehicles at night. An IR variant is available for use with night vision devices, however, chemical lights are not visible through a thermal sight. Lighting systems do not provide for thermal identification during day and will typically not be employed during limited visibility operations (light discipline).

Beacons and Strobes. Beacons and strobes are unit-procured, small-scale, compact, battery-operated flashing devices that operate in the near infrared wavelength. They are clearly visibly through night vision optics, but cannot be viewed through thermal optics. These work both ways, so a near peer adversary with NV capability will see these literally miles away. You should tailor use of the beacon based on METT-TC.

Symbols. Unit symbols may be used to mark friendly vehicles. An inverted V, dots, slashes (or a "Z" as we have seen) for example, painted on the flanks, rear, and fronts of a vehicle, aid in identifying a target as friendly (or enemy). These are typically applied in an area of operations and not during training. Symbol marking systems do not provide for thermal identification during daylight or limited visibility operations. These are easily hijacked and can be used for deception especially when using similar types of equipment.

Glossary - Terms, Acronyms, Abbreviations, and a few limited Definitions

ACE
Ammo, Casualties, Equipment (sometimes reported as LACE, L being Liquids)

Ate Up
Not performing well, unsatisfactory

BMNT
Begin Morning Nautical Twilight

Buddy Team
Two (or three if there is an odd man out) inseparable battle buddies. Implemented for accountability, buddy team rushes/maneuver, and well-being of the unit.

CASEVAC
Casualty evacuation

CM
Citizen Manual

Combat Ineffective
The status of a unit or individual when they become physically or mentally unable to perform assigned warfighting tasks

COTS
Commercial Off The Shelf; referring to an item that is available for free market purchase

CPR
Cardiopulmonary resuscitation

Destruction Plan
Pre determined actions to prevent the imminent capture of sensitive friendly equipment or information

EENT
End Evening Nautical Twilight

FM
field manual

Fire Team
A group of fighters led by a Team Leader (TL), can be three or four individuals with assigned roles and potentially various weapon systems/capabilities (doctrine is four consisting of TL, Automatic Rifleman, Grenadier and Rifleman)

Fire Team Leader
Leads the smallest maneuver element, typically leads three individuals but can be up to 5

FMC
Fully Mission Capable

FRAGO
Fragmentary Order (this change has been determined as unnecessary, so we still use FRAGO)

IFAK
improved first aid kit

Infiltration
A form of maneuver in which an attacking force conducts undetected movement through or into an area occupied by enemy forces to occupy a position of advantage behind those enemy positions while exposing only small elements to enemy defensive fires

IPB Intelligence preparation of the battlefield/battlespace
A systematic process of analyzing the mission variables of enemy, terrain, weather, and civil considerations in an area of interest to determine their effect on operations

Kilometer, K, Click
1,000 meters

LMG
Light Machine Gun

MAG
Mutual Assistance Group

MOS
Military occupational specialty

NLT
Not Later Than

NPA
Nasopharyngeal airway

NMC
Non-Mission Capable (see "Ate Up")

OPORD
Operation Order

SOP
Standard Operating Procedure

STANAG
standardization agreement (NATO)

TC3
tactical combat casualty care

WARNO

Warning Order (current doctrine is "WARNORD", this change has been determined as unnecessary, so we still use WARNO)

WP

white phosphorus (aka "Willie Pete" refers to smoke or weapon), also means white phosphor in context (referring to night vision)

Team

A group working together for a common goal

SPOTREP

Spot Report, use SALUTE format

Squad

A group of fighters led by a Squad Leader (SL), can be seven to thirteen individuals organized into two or three fire teams with assigned roles and potentially various weapon systems/capabilities (Army doctrine is nine men consisting of a SL, 2x TL, 2x Automatic Rifleman, 2x Grenadier and 2x Rifleman)

X Hour

Term we use to designate the moment when a crisis event kicks off

SPOT REPORT (SPOTREP, SALUTE Report format)

This is sometimes referred to as a "BLUE 1" Report. We do not recommend adopting this naming system, it is not needed for our purposes, just something to be aware of. Lines can be used (Alpha, Bravo etc) but that is up to your group and local SOP.

Line ALPHA: Observer or source (omit if it is the calling station, otherwise use call signs or description).

Line BRAVO: Activity or characteristic observed. Use the SALUTE format:

Size: The number of sighted personnel, vehicles, or other equipment.
Activity: What the threat is doing.
Location: Grid coordinates. Report the center of mass for identical, closely grouped items,
otherwise, report multiple grid coordinates of traces.
Unit: Patches, signs, or markings.
Time: Time the observed activity occurred.
Equipment: Description or identification of all equipment associated with the activity.

Line CHARLIE: Actions you have taken and personal recommendations. Actions usually involve conducting additional reconnaissance to determine the complete threat situation or recommending and executing a specific course of action.

Line DELTA: Self-authentication (if required by SOP or tactical situation).

SITUATION REPORT (SITREP)
(This is sometimes referred to as a "BLUE 2" Report)
Subordinate units submit a SITREP on the tactical
situation and status to their leadership or command post.
Submit the SITREP daily (follow your SOP), after
significant events, or when the leadership requests it. State
SITREP followed by pertinent information on these lines:

Line 1: The as-of date-time group (DTG).

Line 2: Brief summary of threat activity, casualties
inflicted, and prisoners captured.

Line 3: Friendly locations (encoded, follow unit SOP)

Line 4: Number of operational vehicles (if applicable).

Line 5: Defensive (friendly) obstacles (encoded using
codes, control measures, or TIRS points). These may
include debris or abatis roadblocks you have emplaced,
wire obstacles etc.

Line 6: Personnel strength classified using the following
status levels:
   GREEN: Full strength; 90% or more fit for combat.
   AMBER: Reduced strength; 80% to 89% fit for
combat.
   RED: Reduced strength; 60% to 79% fit for combat;
the unit is mission capable.
   BLACK: Reduced strength; 59% or less fit for
combat.

Line 7: Water, Ammo, and food supplies available. Status
levels for ammunition and etc are the same ones used for
personnel strength (GREEN, AMBER, RED, or BLACK)
with percentages referring to the amount of basic load level
available. (Refer to Line 6 of this report.)

Line 8: Summary of tactical intentions.

OBSTACLE REPORT (also referred to as a BLUE 9 report).

This is used to report threat or unknown emplaced obstacles to other friendly units or your leadership. Report all pertinent information using the following format:

Line ALPHA: Type of obstacle or obstruction.

Line BRAVO: Location, using grid coordinates. For large, complex obstacles, send the coordinates of the ends and all turn points.

Line CHARLIE: Dimensions and orientation.

Line DELTA: Composition.

Line ECHO: Threat weapons influencing obstacle.

Line FOXTROT: Observer's actions.

Classes of Supply
You may hear these referenced in the community, get familiar with what each denotes as they will appear in military manuals. For example "what is your Class Five status?" is asking what the ammo status is in your unit.

Class I
Subsistence, water, and gratuitous health and comfort items

Class II
Clothing, individual equipment, tentage, organizational tool sets and kits, hand tools, unclassified maps, administrative and housekeeping supplies and equipment, and chemical, biological, radiological, and nuclear (CBRN) equipment

Class III
Petroleum (bulk and packaged), oils, and lubricants

Class IV
Construction fortification and barrier materiel

Class V
Ammunition

Class VI
Personal demand items normally sold through exchanges

Class VII
Major end items

Class VIII
Medical materiel, including repair parts peculiar to medical equipment

Class IX
Repair parts and components

References

ADRP 1-02. *Terms and Military Symbols*, 7 December 2015.

JP 1-02. *Department of Defense Dictionary of Military and Associated Terms*, 8 November 2010.

ADP 5-0. *The Operations Process*, 17 May 2012.

ADP 6-0. *Mission Command*, 17 May 2012.

ADRP 3-0. *Unified Land Operations*, 16 May 2012.

ADRP 3-07. *Stability*, 31 August 2012.

ADRP 3-09. *Fires*, 31 August 2012.

ADRP 3-90. *Offense and Defense*, 31 August 2012.

ADRP 5-0. *The Operations Process*, 17 May 2012.

ADRP 6-0. *Mission Command*, 17 May 2012.

ATP 2-01. *Plan Requirements and Assess Collection*, 19 August 2014.

ATP 2-01.3. *Intelligence Preparation of the Battlefield/Battlespace*, 10 November 2014.

ATP 3-20.98. *Reconnaissance Platoon*, 5 April 2013.

ATP 3-34.81. *Engineer Reconnaissance*, 1 March 2016.

ATP 3-53.2. *Military Information in Conventional Operations*, 7 August 2015.

FM 3-90-2. *Reconnaissance, Security, and Tactical Enabling Tasks Volume 2*, 22 March 2013.

FM 3-98. *Reconnaissance and Security Operations*, 1 July 2015.

FM 4-95. *Logistics Operations*, 1 April 2014.

FM 6-0. *Commander and Staff Organization and Operations*, 5 May 2014.

JP 3-0. *Joint Operations*, 11 August 2011.

JP 3-05. *Special Operations*, 16 July 2014.

www.ingramcontent.com/pod-product-compliance
Lightning Source LLC
Chambersburg PA
CBHW070308200326
41518CB00010B/1932